Geology of the country around Southend and Foulness

This memoir describes the geology of south-east Essex which is bounded by the Thames Estuary to the south and Maplin Sands to the south-east. The dominantly clay upland areas are underlain by Tertiary deposits which have been investigated in several cored boreholes. The reclaimed coastal marshlands of Canvey Island and Foulness Island overlie complex sequences of Flandrian marine and estuarine sediments and the results of various exploratory drilling programmes are synthesised here. On the flanks of the estuaries of the River Thames, Crouch and Roach there are a variety of drift deposits which date back to a time when the Rivers Thames and Medway had a common distributary flowing north-eastwards across the eastern part of this peninsular area.

© *Crown copyright 1986*

First published 1986

ISBN 0 11 884391 5

Bibliographical reference

LAKE, R. D., ELLISON, R. A., HENSON, M. R. and CONWAY, B. W. 1986. Geology of the country around Southend and Foulness. *Mem. Br. Geol. Surv.*, Sheets 258 and 259.

Authors

R. D. Lake, MA, R. A. Ellison, BSc
British Geological Survey, Keyworth, Nottingham NG12 5GG

M. R. HENSON, PhD
Conoco (UK) Ltd, 116 Park Street, London W1

B. W. Conway
formerly *British Geological Survey*

Contributors

J. L. Farr, MSc
Wellfield Consulting Services, P.O. Box 1045, Gaborone, Botswana

G. W. Green, MA
formerly *British Geological Survey*

S. E. Hollyer, BSc, M. J. Hughes, MSc,
R. J. Merriman, BSc and C. J. Wood, BSc
British Geological Survey, Keyworth

C. R. Bristow, PhD
British Geological Survey, Exeter

A. A. Morter, BSc
British Geological Survey, London

Other publications of the Survey dealing with this district and adjoining districts

BOOKS

Memoirs
Geology of the country around Chelmsford, Sheet 241
Geology of the country around Chatham, Sheet 272
Geology of the country around Faversham, Sheet 273

British Regional Geology
London and Thames valley (3rd edition)

MAPS

1:625 000
Solid geology (South sheet)
Quaternary geology (South sheet)

1:50 000 (Solid and Drift)

Sheet 240	Epping
Sheet 241	Chelmsford
Sheet 257	Romford
Sheet 258/259	Southend and Foulness
Sheet 271	Dartford
Sheet 272	Chatham
Sheet 273	Faversham

Produced in the United Kingdom for HMSO

BRITISH GEOLOGICAL SURVEY

R. D. LAKE,
R. A. ELLISON,
M. R. HENSON and
B. W. CONWAY

CONTRIBUTORS

Hydrogeology
J. L. Farr

Palaeontology
A. .A. Morter, M. J. Hughes,
C. J. Wood

Petrology
R. J. Merriman

Stratigraphy
C. R. Bristow, G. W. Green,
S. E. Hollyer

Geology of the country around Southend and Foulness

Memoir for 1:50 000 sheets 258 and 259, New Series

C000821943

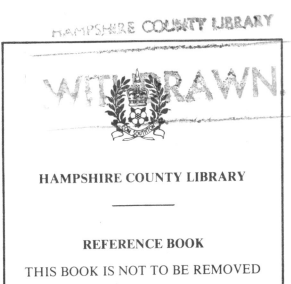
Natural Environment Research Council

LONDON: HER MAJESTY'S STATIONERY OFFICE   1986

CONTENTS

FIGURES

TABLES

PREFACE

This memoir describes the geology of the district covered by the Southend and Foulness (258/259) New Series Sheet of the 1:50 000 Geological Map of England and Wales. The solid geology of the district was originally surveyed on the scale of one inch to one mile by H. W. Bristow and W. Boyd Dawkins and the results were published on Old Series Sheets 1 and 2 (Solid) in 1868. The drift deposits were subsequently surveyed by H. W. Bristow, W. Boyd Dawkins and H. B. Woodward (for Sheet 1) and by H. B. Woodward and F. J. Bennet (for Sheet 2 – revised Solid and Drift) and were published in 1871 and 1883 respectively. Some details of the solid geology of the area were included by W. Whitaker in the memoirs of the London area, published in 1872 and 1889.

A small area in the west of New Series Sheets 258/259 was surveyed on the scale of six inches to one mile by H. G. Dines in 1923 and the entire district was mapped on that scale in 1968 and 1971-72 by Dr C. R. Bristow, Mr G. W. Green, Dr M.R. Henson and Mr R. D. Lake, under the supervision of Mr S. C. A. Holmes, District Geologist. The new 1:50 000 map was published in 1976.

The later part of the geological survey was carried out, in conjunction with research by other divisions of the then Institute of Geological Sciences, in order to assist in the planning of the South-East Essex Development Area. Funds for this work were provided by the Department of the Environment. This volume incorporates many of the results of this extensive geological and geotechnical survey.

Most of this present memoir has been written by Messrs R. D. Lake and R. A. Ellison who have also been responsible for its compilation. Dr M. R. Henson and Mr B. W. Conway have written the chapter on the estuarine deposits. Mr C. J. Wood has supplied details of the Upper Chalk in the Stanford le Hope Borehole. Mr M. J. Hughes, working in conjunction with Mr C. King of Paleoservices Ltd., has contributed on the faunal assemblages of the Tertiary formations. He has also supplied details of the Quaternary microfauna. The results of petrographical analyses, mainly by Mr R. J. Merriman, have been incorporated in the text. Mr S. E. Hollyer and Mr J. L. Farr have provided contributions on the bulk aggregates and water-supply, respectively. The memoir has been edited by Mr W. B. Evans, Dr W. A. Read and Mr G. W. Green.

Grateful acknowledgement is made to numerous organisations and individuals who have generously supplied borehole records and to the landowners who kindly allowed access to their ground.

G. Innes Lumsden, FRSE
Director

British Geological Survey
Keyworth
Nottingham   NG12 5GG

11th April 1986

CHAPTER 1

Introduction

The Southend and Foulness district is bounded by the Thames Estuary to the south and Maplin Sands to the south-east. The western limits of this part of Essex are defined by a line between Stanford le Hope and Stock; a line passing from Stock to just north of Burnham-on-Crouch delineates the northern limits of the district (Figure 1).

Included within the district is the area that was designated for the urban development which was to accompany the construction of the third London Airport at Maplin. The geological survey of the district was initiated primarily in order to assist in the broad planning of the South-east Essex Development Area. Subsequently members of the East Anglia and South-east England field staff, the Mineral Assessment Unit and the Engineering Geology Unit of the Institute conducted an extensive multidisciplinary geological and geotechnical survey of the area. Many of the results of this project are incorporated into this memoir.

The district can be divided into six physiographic regions: the low-lying reclaimed marshlands around Canvey Island in the south-west; a similar marshland area around Foulness Island to the east; the broad valley of the River Crouch; the dominantly clay uplands to the north and to the south of this valley; and an area of broadly undulating ground between Rochford and Shoeburyness. This last area, which is underlain by loams and gravels, provides the best-drained soils, whereas the well-drained former marshland between Burnham-on-Crouch and Shoeburyness is ideal for extensive grain cultivation.

Although the population of the northern part of the district is basically agricultural, the southern part includes several large conurbations which serve as dormitory towns and which also have industrial developments. Tourist and recreational facilities are an important aspect of the towns and villages which are situated close to the Crouch and Thames estuaries.

This part of Essex is drained by numerous small streams which discharge into the sea at Foulness Point by way of the Rivers Crouch and Roach. In the south several small streams drain a limited area around Pitsea and discharge into the Thames Estuary at Canvey Island. In the extreme north-west of the district the drainage of the area is directed northwards to the River Chelmer.

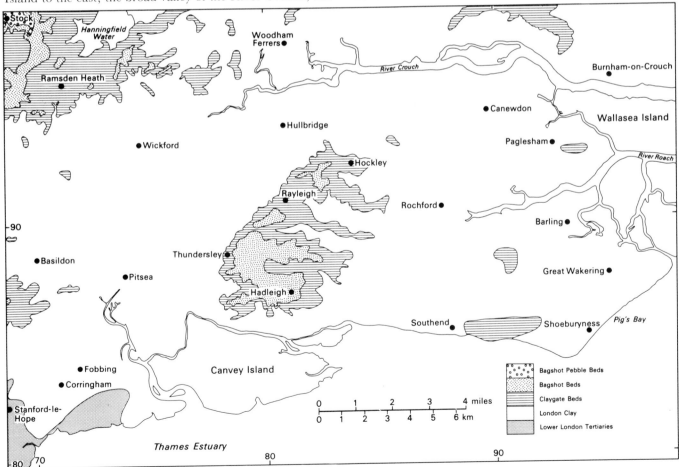

Figure 1   The solid geology of the Southend area

GEOLOGICAL HISTORY

The geological formations represented on the 1:50 000 geological map and sections are summarised on the inside front cover Figures 1 and 2 summarise the solid and drift geology of the district, respectively.

The basement rocks beneath the district form part of the Palaeozoic London-Brabant Massif; they have been recorded in only two bores, at Canvey and Fobbing. The former proved Lower Devonian (Lower Old Red Sandstone) sediments, and the latter proved shallow-water marine sediments, possibly of Lower Carboniferous age. Regional evidence indicates that the district was subject to little subsidence or relative uplift, for a considerable period of geological time; such sediments as were laid down during the interval spanned by the Middle Devonian to Lower Cretaceous have been removed by subsequent uplift and erosion.

A great change occurred during Cretaceous times with a major transgression, which was heralded by the deposition of the Lower Greensand. Deepening of the sea permitted deposition of the Gault clay, but earth movements and subsequent erosion caused removal of the Lower Gault in the present district. The Upper Gault was succeeded by the Chalk which was deposited under marine continental shelf conditions with an extremely restricted influx of terrigenous sediment.

Uplift and mild deformation occurred at the end of Cretaceous times so that the Tertiary sediments rest with slight unconformity on the underlying Chalk. The Lower London Tertiaries (Thanet Beds and Woolwich and Reading Beds) were deposited mainly under shallow marine and estuarine conditions, possibly with a short incursion of fluviatile conditions when the middle Woolwich Beds were deposited. The succeeding London Clay indicates a change to a deeper water and more fully marine environment. Subsequent shallowing of the water is reflected by the two succeeding formations, the Claygate Beds and Bagshot Beds, which become progressively more arenaceous.

The Southend district is peripheral to the North Sea Basin in which thick marine sediments accumulated throughout Mesozoic and Cenozoic time. Certain events, more particularly the earth movements, can be recognised in both areas. Widespread volcanicity occurred in Lower Eocene times in the North Sea Basin and there is evidence to suggest that ash showers may have extended as far as the present district.

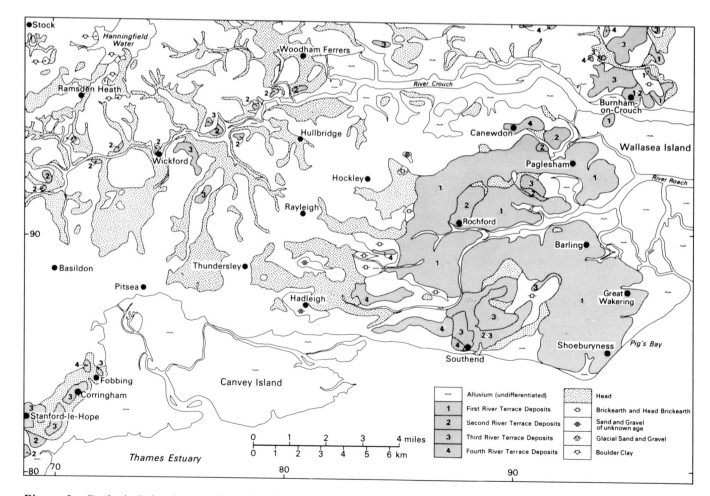

Figure 2   Geological sketch map of the Southend area showing the distribution of drift deposits

Evidence of later Tertiary and early Pleistocene deposits is lacking, partly because of the effects of the Miocene earth movements, which caused significant flexuring of the strata, uplift and subsequent erosion, and partly because of the effects of the Pleistocene glaciations.

The present drainage pattern of the area was developed in post-Tertiary times and was subsequently modified by Pleistocene glacial and periglacial events. The oldest drift deposits belong to the Chalky Boulder Clay (Anglian) glaciation. These were laid down on a pre-existing landscape, much of which was higher than the present land surface. The main Anglian ice-sheet was impounded by the Danbury–Tiptree ridge to the north of the district which restricted incursions of ice from this direction so that deposits of boulder clay and sand and gravel outwash are not widespread. There is little evidence to suggest the presence of sea-ice in the offshore and Thames Estuary areas. When the ice-sheet or sheets receded, the present erosional cycle commenced and only isolated outcrops of glacial drift now remain.

The age of the deposits filling onshore channels near Rochford, Burnham-on-Crouch and Shoeburyness is not precisely known. The evidence indicates that they pre-date the river gravels of the Third Terrace of the Crouch valley system and post-date the glaciation. They accumulated in a restricted, possibly lagoonal, environment under cool climatic conditions, but the channels themselves were probably cut during cold episodes of the Pleistocene.

The Pleistocene period was characterised by a succession of climatic fluctuations which ranged from temperate to fully glacial, and were associated with corresponding changes of sea-level. High sea-levels occurred during temperate phases and low sea-levels during cold phases when much water was removed from the sea in the form of ice. Structural movements in the earth's crust have continued steadily in the Thames Estuary area over the last 2 to 3 million years, and climatically induced oscillations in sea-level were superimposed on these tectonic changes. In the tidal reaches of river valleys, oscillating sea-levels have resulted in the development of terraces and buried channels. During periods of low sea-level, valleys were subjected to erosion whilst the river regraded its profile to the lower base-level. During periods of high sea-level aggradation took place as the river accommodated its profile to the higher base-level. Such changes are reflected in the present district.

Permafrost conditions have subsequently affected all the terraces and resulted in cryoturbation (frost-heave) structures and solifluxion deposits; windblown silt (loess) may have made a substantial contribution to the later Quaternary deposits.

The bases of the various post-terrace alluvial deposits of the rivers and of the Thames Estuary are graded to sea-levels lower than that now prevailing, and relatively recent submergence (since 10 000 years BP) has caused the filling of eroded hollows with gravels, alluvial mud and silt, and the accumulation of similar deposits in the present marshlands.

The drainage system, which was finally established in pre-Roman times after this latest major marine incursion, has been considerably modified by man. Extensive reclamation of marshland around Canvey and Foulness has restricted the rivers and the estuarine tidal flow to well-defined channels.

HISTORY OF RESEARCH

Although no fully comprehensive account of the geology of this area has previously been compiled, several authors have described important aspects of the various deposits in more general texts. Prestwich (1854a; 1854b) first subdivided the Lower London Tertiary sequence, and Whitaker (1872; 1889) described the Tertiary strata of the district. The stratigraphy of the Woolwich and Reading Beds has been described by Hester (1965). The Bagshot Beds of Essex were discussed by Wooldridge (1924), and Curry (1965) summarised the knowledge of the Palaeogene strata of south-east England. The coastal marshlands and offshore Quaternary features have been the subject of intensive studies by D'Olier (1972), Gramolt (1960), and Greensmith and Tucker (1969a; 1969b; 1971a; 1971b; 1973). The Pleistocene gravels have been the subject of interest for many years. Writers on this topic include Gregory (1915), Wooldridge (1923) and Gruhn, Bryan and Moss (1974). The structure of the area was described by Wooldridge and Linton (1955) in their synthesis of the structure and geomorphology of south-east England. The water-supply of Essex was detailed by Whitaker and Thresh (1916). Supplementary well information was provided by the Geological Survey Wartime Pamphlet (Buchan and others, 1940).   RDL

CHAPTER 2

Concealed formations

PALAEOZOIC

Within the present district two deep borings, at Fobbing and Canvey Island, have penetrated rocks of Palaeozoic age. The Canvey Island Borehole [8215 8330][1] proved 130.9 m of Palaeozoic sandstones with subordinate conglomerates, mudstones and siltstones in variegated shades of red, brown, green and grey (Smart, Sabine and Bullerwell, 1964). Palynological evidence indicates an Emsian (Lower Devonian) age for these deposits (Mortimer, 1967). The Fobbing Well [7127 8436] penetrated 7.8 m of Palaeozoic shales, sandstones and quartzites, which were ascribed first to the Cambrian (Dewey, Pringle and Chatwin, 1925) and later, tentatively, to the Lower Carboniferous (Stubblefield *in* Trueman, 1954, p.163). Evidence from these and other boreholes in the surrounding area (See Table 1 and Figure 3) indicates that the Paleozoic floor lies at depths between −300 m and −500 m OD in the Southend district; this surface is thought to slope gently down eastwards towards the North Sea Basin.

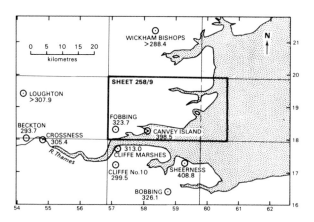

Figure 3   Location map of deep boreholes in the Thames Estuary region

Table 1   Summarised logs of deep boreholes in the Thames Estuary area

	BGS Well Catalogue number	Ground level (m OD)	Thickness (m)			Depth of Palaeozoic floor and formation
			Drift and Tertiary	Chalk	Upper Greensand and Gault	
Beckton	257/39h	3.8	39.0	197.2	61.0	− 293.7 Devonian (Emsian)
Crossness	257/3	1.8	41.8	192.3	73.1	− 305.4 ?Devonian
Loughton	257/82	27.4	74.1	198.4	61.5 +	
Wickham Bishops	241/58	71.3	145.3	214.3 +		
Fobbing*	258/42b	19.8	91.0	191.6	51.2	− 323.7 ?Carboniferous
Canvey Island	258/148	3.0	157.0	206.4	38.3	− 398.5 Devonian (Emsian)
Cliffe Marshes†	272/23b	3.7	23.5	199.9	63.4	− 313.0 Silurian
Bobbing‡	272/24	36.6	43.3	207.3	54.3	− 326.1 Silurian
Sheerness**	272/90	2.1	153.0	203.6	40.9	− 408.8 Silurian
Cliffe No. 10***	272/274	2.5	36.0	205.4	56.4	− 299.5 ?Devonian

* See discussions below on the Lower Greensand occurrence.
† This borehole additionally proved 29.9 m of Lower Greensand beneath the Gault.
‡ This borehole additionally proved 57.9 m of Lower Greensand and Jurassic strata beneath the Gault.
** This borehole additionally proved 13.4 m of Lower Greensand beneath the Gault.
*** This borehole additionally proved 4.26 m of Lower Greensand beneath the Gault.

1 National Grid References are given in this form throughout. Unless otherwise stated figures with eastings between 6500 and 9999 relate to places in 100-km square TQ (or 51), those with eastings between 0000 and 0610 to places in 100-km square TR (or 61).

CRETACEOUS

Strata of Jurassic age are unlikely to be preserved beneath the Cretaceous rocks of the Southend and Foulness district, although a localised occurrence within a fault-bounded graben has been proved at Cliffe to the south of the district (Owen, 1971).

The Fobbing Well (Dewey and others, 1925, p.133) proved the following Cretaceous strata:

		Thickness m	Depth m
(v)	Chalk (see below)	187.0	278.0
(iv)	Glauconitic Marl (basal Lower Chalk)	4.6	282.6
(iii)	Calcareous sandstone slightly glauconitic	5.4	288.0
(ii)	Dark clay (Gault)	45.7	333.7
(i)	Whitish quartz sand with dark shale and grit—probably a conglomerate (?Lower Greensand)	9.8	343.5

Bed (iii) was assigned by these authors to the Upper Greensand. They grouped (i) with the Gault, but subsequent writers (Kirkaldy, 1933, p.300; Owen, 1971, p.197) have taken this unit to be Lower Greensand and have suggested that the Lower Gault is absent.

A comparable sequence proved in the Canvey Island Borehole (Smart and others, 1964, p.18) was classified as follows:

	Thickness m	Depth m
Upper Chalk	84.7	241.7
Middle Chalk	68.3	310.0
Lower Chalk	53.3	354.2
Glauconitic Marl, sandy in lower 3.0 m (basal Lower Chalk)	9.1	363.3
Calcareous, micaceous sandstone (Upper Greensand) (see below for revised classification)	4.3	367.6
Dark greenish grey marl (Upper Gault)	34.0	401.6

Here again Lower Gault was absent.

These boreholes, taken in conjunction with other nearby penetrations, indicate that the Upper Greensand is absent to the south and south-east of Canvey Island and that the Lower Gault is absent over much of southern East Anglia as a result of Middle Albian tectonic warping and subsequent erosion. The Lower Greensand has a restricted occurrence and probably occupies an embayment in the Fobbing and Cliffe area.

The Chalk proved in the Canvey Island Borehole was subdivided on the evidence of electric logs (Smart and others, 1964, p.4). The occurrence of the Plenus Marls at the top of the Lower Chalk provided confirmatory lithological evidence of the Lower–Middle boundary. Generally, however, the division between the Upper and Middle Chalk, which are both flint-bearing, is poorly defined in percussion drilled boreholes because the Chalk Rock, which marks the base of the Upper Chalk, is difficult to detect. In the Fobbing Well 152.9 m of Upper and Middle Chalk were thought to overlie 38.7 m of Lower Chalk. This subdivision depended on the observation of 'green ?marl' at a depth of 243.8 m and, in the light of the evidence from the Canvey Island Borehole, the correlation of this band with the Plenus Marls may be erroneous.

Table 1 shows the thicknesses of Chalk proved in the neighbouring areas. RDL

Details

LOWER CRETACEOUS   The bed numbers from the scheme devised for East Anglia by Gallois and Morter (1982) may be applied to the Gault succession of the Canvey Island Borehole. The sequence is:

		Thickness m	Depth m
Lower Chalk	to	—	363.32
Upper Greensand (?Stoliczkaia dispar Zone) Bed 18 of Gallois and Morter recognised below 365.76 m		4.88	368.20
Upper Gault (Mortoniceras rostratum Subzone) Bed 17, rich in Aucellina		0.71	368.91
(Callihoplites auritus Subzone) Bed 16		15.54	384.45
Bed 15 (Hysteroceras varicosum Subzone)		6.79	391.24
Bed 14 (Hysteroceras orbignyi Subzone)		8.81	400.05
Bed 13		1.40	401.45
?Bed 12		0.12	401.57
Lower Devonian		—	—

Bed 17 is the equivalent of Bed XII of the Folkestone succession, which is now taken as the base of the dispar Zone (see Gallois and Morter, 1982). The lower part of the Upper Greensand probably lies within the rostratum Subzone although faunal evidence is lacking. A study of the eiffelithid nannofossils by Dr A. W. Medd has confirmed this zonation. AAM

UPPER CRETACEOUS   The fauna from the 1.51 m of Upper Chalk proved in the Stanford le Hope Borehole [6965 8241] includes the following:

lituolid foraminifera; Porosphaera?; Parasmilia granulata Duncan; Neomicrorbis sp. [loosely coiled form, possibly late growth-stage of N. crenatostriatus (Münster)], Orbirhynchia cf. pisiformis Pettitt; Terebratulina sp. [juv.]; Echinocorys [large test fragment]; Micraster? [test fragment], Stereocidaris cf. sceptrifera (Mantell) [radioles]; asteroid marginals, including Metopaster uncatus (Forbes); Glenotremites sp. [distal brachial]; inoceramids [specifically indeterminate shell pieces and a hinge fragment]; Limea granulata (Nilsson); Pseudoperna boucheroni (Woods non Coquand) [irregularly-shaped broad forms with flat attachment area]; fish-scales, not determined.

The macrofauna is not zonally diagnostic, but is strongly suggestive of the Micraster coranguinum Zone or the overlying Uintacrinus socialis Zone. Parasmilia granulata is recorded from the coranguinum Zone, but is said to range up to the Marsupites testudinarius Zone. The occurrence of the broad flat-based Pseudoperna rather than the narrow subcylindrical forms typical of the crinoid zones points to a *coranguinum* Zone position, as does the diversity of the echinoderm fauna.

The foraminiferal assemblage, however, allows the stratigraphical position of the Chalk succession to be located with considerable accuracy within the Santonian part of the *coranguinum* Zone. Microsamples from 57.10, 58.30 and 58.58 m were examined by Dr H. W. Bailey (Paleoservices Ltd) and also by Dr P. Bigg (formerly of the Palaeontology Unit). The occurrence of *Stensioeina granulata polonica* Witwicka in all three samples indicates the *S. granulata polonica* Foraminiferal Biozone (see Bailey and others, 1983, fig.2 and p.39), corresponding to the Santonian part of the *coranguinum* Zone and the basal part of the *socialis* Zone. Within this relatively broad zone, the stratigraphical position can be narrowed down to the lower part of the *Cibicides* ex gr. *beaumontianus* Foraminiferal Assemblage Biozone (Bailey and others, 1983), on the basis of the occurrence of *Cibicides beaumontianus* (d'Orbigny), *C. beaumontianus* var. A (Bailey MS), *Eponides* cf. *concinna* Brotzen, *Eouvigerina gracilis* (Egger), *Pyramidina buliminoides* (Brotzen) and common *Stensioeina exsculpta exsculpta* (Reuss), together with *Lingulogavelinella* sp. cf. *L. vombensis* (Brotzen) and *Reussella kelleri* Vasilenko. *L.* cf. *vombensis* constitutes over 12 per cent of the fauna in the sample from 57.10 m: in the Kent coast sections, this species increases in frequency upwards, and has not been found higher than about 2 m above the base of the Assemblage Biozone, at approximately the level where *R. kelleri* re-enters the succession following a long gap in its vertical range. In lithostratigraphical terms, the borehole succession can therefore probably be located just above the equivalent of Whitaker's 3″ Flint Band of the Kent succession (see Bailey and others, 1983, fig.2). Both Dr Bailey and Dr Bigg noted the presence of *Conorbina marginata* Brotzen and *'Discorbis' scanicus* Brotzen, two Santonian species described from Eriksdal in Sweden (Brotzen, 1936), but not previously recognised in the UK. A detailed range chart showing all the species present, together with data on percentage occurrences in the individual samples, is deposited with the BGS Biostratigraphy Research Group.   CJW

CHAPTER 3

Tertiary (Palaeocene and Eocene)

PALAEOCENE

Nomenclature and general stratigraphy

The base of the Palaeocene in the London Basin is drawn at an unconformity which represents a period of erosion of the Chalk prior to Tertiary times (Figure 4). The definition of the boundary between the Palaeocene and Eocene Series is currently under debate. The 'Palaeocene' Series was defined by Schimper (1874) primarily on palaeobotanical studies in the Paris Basin; he included the Ypresian Stage within the Palaeocene though he placed the London Clay, with its Ypresian faunas, in the Eocene (see Figure 5). However, it is now accepted that the Ypresian Stage is better placed in the Eocene, and accordingly workers in Britain (Curry, 1966;

Cooper, 1976; Fitch, Hooker, Miller and Brereton, 1978) have placed the upper boundary of the Palaeocene at the top of the Oldhaven Beds of Kent which is at the Sparnacian–Ypresian boundary. In the present account, as on the published map, the top of the Palaeocene has been drawn at a horizon that corresponds closely with that of the top of the Oldhaven Beds of Kent, namely at the base of the strata mapped as London Clay in the present district. Nevertheless, it should be remembered that Knox (1984), who has correlated evidence from nannoplankton assemblages and ash bands, would prefer to place the Palaeocene–Eocene boundary within the Woolwich and Reading Beds; within the present district this would not correspond to any easily mapped boundary.

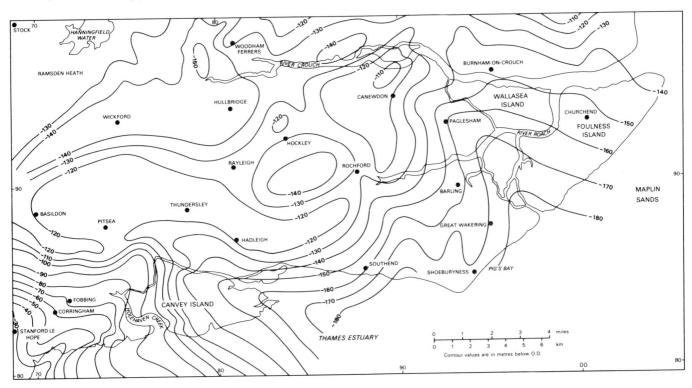

Figure 4   Contours on the base of the Tertiary deposits in metres below OD

Figure 5
Schematic sub-
division of the
Palaeocene strata
to indicate terms
used in this
memoir

SOUTHEND AND FOULNESS (258/259)
1:50 000 GEOLOGICAL SHEET

		LITHOSTRATIGRAPHY		STAGE	PERIOD
LOWER LONDON TERTIARIES	WOOLWICH BEDS INCLUDING OLDHAVEN BEDS	OLDHAVEN BEDS, INCLUDING LONDON CLAY BASEMENT BED		SPARNACIAN	PALAEOCENE
		READING FACIES			
		WOOLWICH FACIES			
		WOOLWICH BOTTOM BED FACIES			
	THANET BEDS	THANET BEDS		THANETIAN	

In the London Basin, the Palaeocene rocks comprise sands, conglomerates and subordinate clays that have long been collectively called the Lower London Tertiaries. They were first formally subdivided by Prestwich (1850; 1852; 1854a, b,) who recognised the 'Thanet Sands', the 'Woolwich and Reading Series' and the 'Basement Bed of the London Clay'. These terms cover strata now generally known respectively as the Thanet Beds, Woolwich and Reading Beds and the Oldhaven Beds, whose generalised stratigraphy in the eastern part of the London Basin is shown in Figure 6. Cooper (1976) formalised the Lower Tertiary nomenclature; he referred to the Lower London Tertiary Group, and divided this into the Thanet, Woolwich and Reading, and Oldhaven formations. The base of the Oldhaven Beds is now known to mark the onset of a major marine transgression and to be diachronous on a regional scale (Knox, Harland and King, 1983) so that strictly these beds are best excluded from the Lower Tertiary Group and classified as a basal sand member of the London Clay Formation. However, within the present district, the Woolwich and Reading Beds and the Oldhaven Beds are not readily separable. Thus they have had to be grouped together on the 1:50 000 sheet and in the following account.

The Thanet Beds are typically pale greenish buff, fine-grained, bioturbated, glauconitic, marine sands, which may locally contain sporadic dark clays. The term Thanet Beds has not changed in stratigraphical usage since its inception and their lithology is too uniform in this district to warrant any useful subdivision. In contrast, the Woolwich and Reading Beds, a term first used by Whitaker (in Hull and Whitaker, 1861), vary in lithology from pebble beds to clays, and were deposited in environments ranging from marine to freshwater (Ellison, 1983). Hester (1965) subdivided them into the Reading Beds, the Woolwich Beds, and the Bottom Bed, but these variants are now regarded as facies, which are here termed the Reading facies, the Woolwich facies, and the Bottom Bed facies (Figure 5). The first is present only in the extreme north and, locally, in the east of the district, whereas the Woolwich facies and the Bottom Bed facies are found over the whole district. To conform with the published 1:50 000 Southend (258) Sheet, all three facies are termed Woolwich Beds throughout this account, though the unit should more properly be called the Woolwich and Reading Beds. Ellison (1983) has demonstrated how the facies represented in the Woolwich and Reading Beds of this area fit into the overall pattern of facies variations within the London Basin as a whole.

Over the years, there has been some confusion in the nomenclature of the marine sands and clayey sands that lie between the Woolwich and Reading Beds and the base of the typical marine clays of the London Clay. Whitaker (1866; 1872) subdivided the 'Basement Bed of the London Clay' into the Oldhaven Beds and the 'London Clay Basement Bed', naming the former after their best exposure in Oldhaven (Bishopstone) Gap near Herne Bay, Kent. Within the Southend district the London Clay Basement Bed and the Oldhaven Beds are lithologically and faunally indistinguishable, so the latter name has been used in the present account for the resulting combined lithostratigraphical unit. On a regional scale this sandy facies at the base of the London Clay is known to be diachronous (Knox and others,

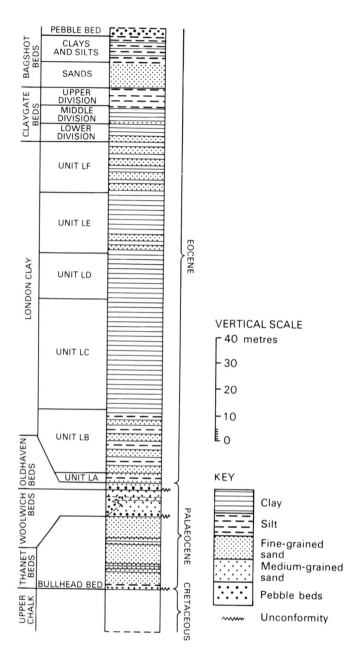

Figure 6 Generalised vertical section of Tertiary beds in south-east Essex

1983).

In south-east Essex the Lower London Tertiaries, including the Oldhaven Beds, are mainly sands, which vary in thickness from less than 35 m beneath the Dengie peninsula in the north-east to more than 50 m adjacent to the River Thames in the south (see Figure 7). They occur at depth throughout the present district, but crop out only west of Stanford le Hope where they have been mapped as Woolwich Beds (including Oldhaven Beds).

The BGS cored boreholes, at Stanford le Hope[1] [6965

---

1 The boreholes at Rainbow Lane, Stanford le Hope, and at Sandpit Hill, Hadleigh, are subsequently referred to in this account as the Stanford le Hope and Hadleigh boreholes respectively (see Appendix 2).

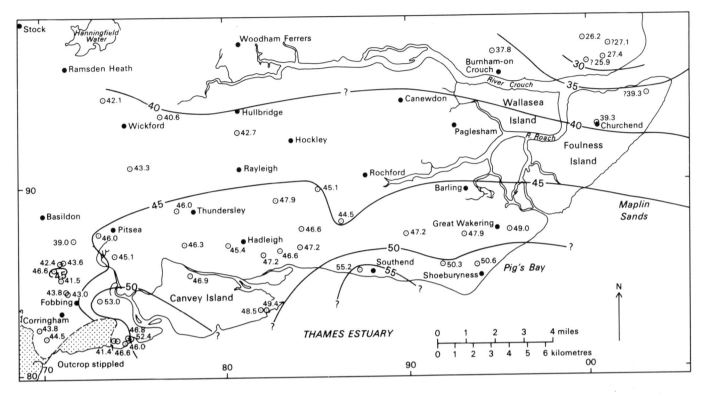

Figure 7   Generalised isopachyte map of the Lower London Tertiaries

8241] and Hadleigh [8002 8654], were drilled to supplement the ground survey, and further detailed information about lithologies and facies relationships has been obtained from temporary and quarry sections in the Stanford le Hope, Orsett and Linford areas of the adjoining Romford (257) district; this is recorded below. The Lower London Tertiaries have been penetrated by a large number of water boreholes and wells (Whitaker and Thresh, 1916 and BGS unpublished records) which have provided further information regarding their regional variation though, because of the predominance of sand, drillers' logs in South Essex cannot be interpreted in detail. In particular, the complicated facies variations within the Woolwich Beds and Oldhaven Beds are difficult to unravel with certainty—a difficulty which is increased by the variable degree to which the sediments have been reworked during deposition.

Thanet Beds

In the present area the Thanet Beds lie unconformably on the eroded Upper Chalk surface. They do not crop out at the surface but are present immediately beneath the Drift deposits near Mucking.

The basal bed, known as the Bullhead Bed, was first described by Morris (1876); it contains large nodular flints coated with bright green glauconitic clay. In the Stanford le Hope Borehole, the Bullhead Bed was represented by 1 to 2 cm of clayey, glauconitic, medium-grained sand, overlain by 10 cm of nodular 'bullhead' flints up to 6 cm in diameter. Above it lay a uniform, greenish-grey, bioturbated, silty, fine-grained sand which rapidly oxidised to a yellowish grey colour. Small, black, rounded flint pebbles, up to 1 cm maximum diameter, were present in the lowest 30 cm of these

sands. Farther east, a borehole [8628 8804] at Eastwood Pumping Station proved up to 7 m of dark coloured clays within sands. Other boreholes in the Southend area have indicated clays at a similar stratigraphical horizon.

Hester (1965, fig. 7), in a synthesis of all the then available borehole and well information, constructed an isopachyte map which showed that the Thanet Beds are present in Essex, north Kent, Surrey and Suffolk. The thickest deposits, which are at least 33.5 m thick, apparently lie just to the north of the Isle of Sheppey. The thickness of 28 m which was recorded in the Stanford le Hope Borehole is consistent with this isopachyte map which indicates about 32 m of Thanet Beds present beneath Southend, thinning to 25 m in the north and north-east of the present district.

No macrofauna was recovered from the Stanford le Hope Borehole, and pyritised diatoms in the lower part constitute the only microfauna. Pyritised diatoms have also been recorded from 5 to 6 m above the Upper Chalk in the type section of the Thanet Beds at Pegwell Bay, Kent, (Burrows and Holland, 1897, p.29). The microfauna from Stanford le Hope is the first record of diatoms west of Pegwell Bay and may represent a westward extension of this horizon. In the Hadleigh Borehole, a fragmentary molluscan fauna including *Lucina*, *Nucula* and *Nuculana*, together with a sparse foraminiferal fauna containing *Ceratobulimina tuberculata* Brotzen, lay at the top of the Thanet Beds.

Woolwich Beds

In the Southend district, the Woolwich Beds may be distinguished from the Thanet Beds by a marked increase of grain size above a basal pebble band. They comprise predominantly medium-grained sand with a variable glauconite content and with subordinate clay laminae; localised pebble beds and coarse-grained sand or sandstone beds also occur. The Woolwich Beds are found at depth beneath the whole district and are generally about 10 m thick. In terms of Hester's classification (see p.8), the Woolwich Beds of most of this district comprise two main facies: a marine glauconitic, finely laminated, sandy lower division (the Bottom Bed facies) with a thin and probably impersistent basal pebble bed; and an upper division of marine and estuarine sands (Woolwich facies) containing subordinate clayey beds. There is, in addition, a third division comprising mottled clays (the Reading facies). This latter is only locally present although to the north and east of the present district it consistently occurs at the top of the Woolwich and Reading Beds, immediately beneath the sandy beds at the base of the London Clay (Whitaker and Thresh, 1916, p.191; Ellison, 1983; Bristow, 1985).

Details

Good sections in the Woolwich and Reading Beds are exposed a few kilometres west of the district. The basal pebble bed of the Woolwich Beds is visible in a quarry section near Thurrock [6600 7980], 3 km west-south-west of Stanford le Hope, where it comprises a 20 cm bed packed with a small, rounded, black flint pebbles up to 1.5 cm in diameter, set in a medium-grained sand matrix. Variably glauconitic, bioturbated, cross-bedded and planar-bedded, medium-grained sands, with laminae of light and dark grey clay normally up to 3 cm thick, overlie this basal bed. *Ophiomorpha* burrows in these sediments at Orsett Depot [656 810], Orsett Cock Quarry [6560 8116] and Hall's Gravel Pit, Linford [6660 7980] indicate a shallow-water marine environment. Similar sediments are recorded from the Stanford le Hope and Hadleigh Boreholes.

Some of the Woolwich Beds do not display the fine lamination typical of the marine beds in this formation. These exceptions occur in the middle of the sequence and are commonly brown and grey colour-banded, medium- and coarse-grained sands, and silicified purplish sandstones with thin lignitic clay horizons. They may have been deposited in estuarine conditions. At Holford Road Corporation Tip, Linford [6701 8134], in the adjacent Romford (257) district about 2.5 km west of Stanford le Hope, a partially silicified, purplish, medium-grained sand horizon overlies at least 0.6 m of banded, light grey-brown, medium- and coarse-grained sand. In the Stanford le Hope Borehole a similar banded hard sand bed, 2.1 m thick, was overlain by sands containing lignitic horizons 1 cm thick.

Beds of lignite, lignitic clay, and sand 6.4 m thick were also penetrated in a borehole [9346 8506] at Shoeburyness where they directly underlie the Oldhaven Beds. A quartzitic sandstone was recorded at 22 m depth in the Stanford le Hope Borehole, and another was penetrated in the Hadleigh Borehole in the upper 3 m of Woolwich Beds. In the Hadleigh Borehole, the presence of the brackish water bivalve *Corbicula cuneiformis* (J. Sowerby) in the uppermost 3 m of the Woolwich Beds suggests an estuarine environment of deposition.

Only two wells have penetrated beds of Reading facies. At Burnham-on-Crouch a well [9473 9703] penetrated 1.37 m of 'coloured clay' beneath shelly sandy clay at the base of the London Clay, and a well at Eastwood [8334 8986] proved 9.8 m of mottled clay beneath the London Clay.

Oldhaven Beds

The term Oldhaven Beds was first used by Whitaker (1866) for about 5 m of uniform cross-bedded sands containing lenses of glauconitic shelly sand lying above a basal pebble bed of small rounded black flints which rests unconformably on the Woolwich Beds. The sands are appreciably finer in grade than those of the latter formation. The Oldhaven Beds are indistinguishable from the London Clay Basement Bed, described by authors including Whitaker (1866) from other localities in the London Basin; for reasons of practicality both are classified as Oldhaven Beds in this account, and have been grouped with the Woolwich Beds on the 1:50 000 sheet.

The Oldhaven Beds are exposed only in the south-western extremity of the district near Stanford le Hope where partially decalcified indeterminate shells in a sandy matrix were seen in a temporary section [684 833]. Exposures to the west of Stanford le Hope in the Romford (257) district, indicate that a disconformity separates the Oldhaven Beds (including the London Clay Basement Bed) from the London Clay, and that the base of the latter in these localities is usually well defined and gently undulating. Well records, notably south of a line between Basildon, Rochford and Burnham-on-Crouch, commonly record 1.2 to 7 m of shelly and pebbly sands beneath the London Clay. Presumably these are the Oldhaven Beds (see, for example, Whitaker and Thresh 1916, p.175 and p.236). A khaki-coloured shelly glauconitic sand penetrated in the Stanford le Hope Borehole immediately beneath the London Clay is of a similar lithology to the Oldhaven Beds at Herne Bay.   RAE

Details

In the adjacent Romford (257) district a temporary exposure at Pump Lane, Horndon on the Hill [6710 8288], 4 km west of the Stanford le Hope Borehole, showed 0.5 m of glauconitic fine- and medium-grained sand with shelly lenses, immediately beneath the London Clay. A similar section is evident in an old quarry at Holford Road, Thurrock [6701 8134]. Here, the basal part of the London Clay contains glauconitic sand lenses a few centimetres thick, and rests with disconformity on 1.7 m of cross-bedded, yellow-orange, fine-grained sand with light grey clay laminae in the lowest 40 cm. Rounded black flint pebbles occur in the basal 5 cm of this bed.

Dines (*in* MS, 1923) noted the following section beneath third terrace deposits at a gravel pit [6928 8275], now back-filled, near Stanford le Hope school:

Blue-green, lilac and pink clay, rather shaly or bedded	0.3 to 0.4 m
Loamy sand with a layer of small black pebbles near the top	9.0 m

The clay in this section represents the basal London Clay, and the sand and pebble bed beneath probably represents the London Clay Basement Bed or Oldhaven Beds. None of these sections is now open.   RAE, RDL

EOCENE

Nomenclature and general stratigraphy

The Eocene strata in this district comprise a conformable sequence of marine sediments which reach a maximum thickness of approximately 180 m. The lithology varies from fine-grained sands and pebbles to silty clays; the coarser beds occur at the base and top of the sequence (see Figure 6).

Within the eastern part of the London Basin the Eocene Series has been traditionally divided into London Clay, Claygate Beds and Bagshot Beds, which are included in the Ypresian Stage, following the nomenclature adopted by Curry (1958). The beds have been given a formal stratigraphical nomenclature by Cooper (1976), who has termed the entire sequence the London Clay Formation, in which the Claygate Beds and Bagshot Beds have named member status. King (1981) has put forward an alternative formal stratigraphical classification but even this is now known to require modification (Knox and others, 1983). In the present memoir the terms London Clay, Claygate Beds and Bagshot Beds are each used in a formational sense, following the proposals of Bristow, Ellison and Wood (1980, pp. 261 – 262), in order to conform with the terminology of the published 1:50 000 sheet. All three formations are, together with the Oldhaven Beds, essentially facies divisions and, on a regional scale, strata of one formation are known to pass laterally into another (King, 1981, fig. 3; Knox and others, 1983, fig.1).

The London Clay comprises predominantly the clayey sediments that overlie the Oldhaven Beds. Wrigley (1924) subdivided the formation into five on the basis of faunal and lithological variations and, although strictly applicable only to the London area, these divisions have been loosely extended to other parts of the London Basin. A more simplified lithological subdivision into upper, middle and lower divisions was applied by Burnett and Fookes (1974) in a review of the regional variation of the engineering properties of the London Clay. As part of the present survey, a standard local London Clay succession for south-east Essex was established from the Hadleigh Borehole; six lithological and fourteen faunal units were recognised and are considered below. A comparison with the lithological sub-divisions of Wrigley and those of Burnett and Fookes is shown, together with the biofacies units of King (1981), in Figure 8. In the usage followed in this memoir and on the 1:50 000 Sheet, the London Clay does not include the London Clay Basement Bed, which has been grouped with the Oldhaven Beds and described with the Palaeocene strata. The basal part of the London Clay as described in this district is characterised by fine sandy and silty beds, with scattered small pebbles. The lowest 7.92 m of London Clay penetrated in the Stock Borehole[1] was originally classified by Bristow (1971, p.21) as Woolwich and Reading Beds but, because C. King (in Bristow, 1985) regards the fauna from this interval as indicative of the basal London Clay as recognised by him at Walton-on-Naze and Cheshunt (where it can be seen clearly to overlie the basal pebble bed of the London Clay), this part of the sequence has been reclassified. Similar sandy beds in the same stratigraphical position in the Hadleigh Borehole have been classified as London Clay. Further details regarding these beds are to be found on p.14.

The Claygate Beds broadly represent a passage by alternation up from the London Clay into the sands of the Bagshot Beds. They occur in Surrey, north London and in Essex, where they rest conformably on the London Clay; in the pre-

<hr />

1 The IGS Stock Borehole [TL 7054 0045] was drilled in 1971 some 17.5 km north-west of Hadleigh and north of the district described in this memoir, in the adjoining Chelmsford (241) district. The borehole is fully documented by Bristow, 1985.

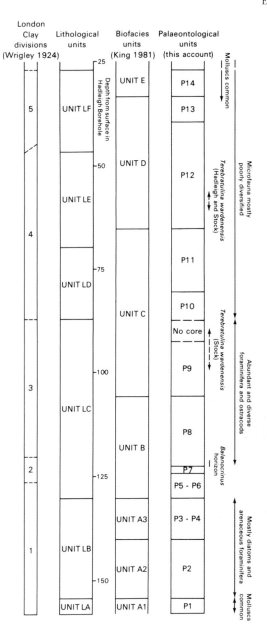

Figure 8 Comparison of lithological and palaeontological subdivisions of the London Clay

sent district three subdivisions have been made on the basis of lithological and palaeontological variation. The term Claygate Beds was given by Dewey (1912, p.239) to the beds between the London Clay and Bagshot Beds exposed around Claygate and Esher in Surrey. He described them as follows: 'They differ from the London Clay in being even-bedded in layers; in containing fine laminae of white sand between the clay beds at the base, and thin seams of clay between the thicker beds of sand in the upper part of the sequence'. Dines and Edmunds (1925) were the first to map the Claygate Beds during the survey of the Romford (257) Sheet. The same strata were called Passage Beds by Wooldridge (1924) and Wooldridge and Berdinner (1925). Wooldridge (1924, p.368) recognised two facies of the Passage Beds in Essex; a structureless loam typical of the Brentwood area, and a

Claygate-type facies found farther south around Vange, Hadleigh and Rayleigh. The former facies, which has probably been bioturbated, is found in the north of the present district around Hanningfield, whereas the Claygate-type facies occurs at all localities recorded south and east of Rettendon. In the present district it is recognised that the base of the Claygate Beds has been drawn at a lower stratigraphical horizon than that taken by Dewey (1912) in the type area (Bristow and others, 1980; Bristow, 1985), and King (1982) has argued that the Claygate Beds should be restricted to strata having the original finely laminated facies of the original type area.

Nevertheless, in the present district the criteria of Bristow and others (1980) have been followed, essentially on the grounds of mappability. An additional factor is that the extended Claygate Beds formation, as mapped, helps to define the areas most liable to landslips (see p.53).

The first reference to 'Bagshot sands' in Essex was by Prestwich (1854b, p.402), and a further account was edited by Whitaker (1889). Subsequent work in Essex by Wooldridge (1924), Berdinner (1925), Wooldridge and Berdinner (1925) and Dines and Edmunds (1925) referred to the Bagshot Beds, which is the term used in this Memoir. A three-fold sub-division, comprising in ascending order a sand division, a clay and silt division, and the Bagshot Pebble Bed, was proved in the Stock Borehole and described fully by Bristow, (1985); it has been adopted for this memoir (see Figure 6). Of the three sub-divisions, the lowermost corresponds to the typical sands of the Bagshot Beds described by earlier authors. The clay and silt sub-division occurs only in the Stock area in the north-west of this district, where Whitaker (1872, p.327) referred to a 'brickearth' overlying the sands of the Bagshot Beds. Pebbly beds associated with the Bagshot Beds were first recognised by Wood (1868, p.464). Subsequently the term Pebble Bed was used by Whitaker (1889) and, in their survey of the Romford area, Dines and Edmunds (1925) formally named the Bagshot Pebble Bed. The term Bagshot Pebble Bed is used in this memoir, although within the district its relationship to the sands of the Bagshot Beds is uncertain. In the London Basin pebble horizons interbedded with undoubted Bagshot Beds have been recorded only at Langtons [TQ 578 948] near South Weald (Whitaker, 1889, p.273; Dines and Edmunds, 1925, p.17), some 14 km south-west of the Stock Borehole.

London Clay

The London Clay is present over the whole of this district except for a small tract in the south-west around Stanford le Hope. It comprises predominantly clayey sediments with sandy beds at its base and its top.

The argillaceous beds of the London Clay are olive to brownish grey in colour, weathering to a chocolate-brown colour owing to the oxidation of their contained pyrite in the zone of water-table fluctuation. The more arenaceous beds, which when unweathered are similar in colour to the argillaceous beds, weather to an orange-brown colour. Blue or bluish green reduction veins are commonly found on joint and fissure surfaces. At outcrop and in boreholes the London Clay generally weathers to a depth of 8 or 9 m but beneath thick gravel deposits this 'weathered zone' may only be a few centimetres thick.

Around the Leigh-on-Sea – Southend conurbation and in the Rochford area, London Clay has been proved in numerous shallow boreholes and temporary exposures which are too numerous to be listed individually. The typical shallow exposure shows completely oxidised brown clay which becomes increasingly stiff downwards and commonly includes lines of septarian nodules. The latter are valuable in shallow exposures for they help to distinguish highly weathered, but still in situ London Clay from reworked Head deposits derived from the London Clay which can locally be similar in appearance. In deeper excavations there is, in places, a zone of patchy oxidation in which oxidised material occurs in the vicinity of joints or as random, irregular patches.

The full thickness of London Clay in this district is known only where the formation is capped by the Claygate Beds, although in a few places it has been possible to calculate its original value from the position of the base of the Claygate Beds in adjacent outcrops. The Hadleigh Borehole proved 132.2 m; other thicknesses quoted below are not precise

NGR	Borehole/Well No.	Thickness m	(the 'full sequence' includes sand at the base of the London Clay)
6774 9510	257/150	125.88	
6895 9305	258/66	109.73 +	(full sequence c.128)
7054 0045	Stock Borehole	126.3	(full sequence >132.5)
7119 9648	258/64	133.5 +	
7161 8718	258/109	120.4 +	
7333 9472	258/31	115.97 +	(full sequence c.136)
7743 9537	258/70	126.18 +	(full sequence c.129)
8002 8654	Hadleigh Borehole[1]	127.93	(full sequence 132.2)
8024 8674	258/50	c.128.00	
8277 8917	258/34	c.136.00	
8923 9878	258/53	132.28	

[1] See Appendix 2

because of the difficulty in defining the upper and lower limits of London Clay in borehole logs. For accurate comparison it must be stressed that the following thicknesses proved in boreholes do not include any sand recorded at the base of the London Clay, except where stated. Many of the logs of these boreholes have been reclassified during the present survey and further details regarding their interpretation are to be found on pages 16 to 17.

Lithological and faunal units

In the following account the London Clay has been subdivided into six lithological units on the basis of the field descriptions of the Hadleigh Borehole which penetrated the full thickness (see Table 2 and Appendix 2). Grain-size analyses to ½-phi intervals, which were subsequently carried out on random samples, proved the basic utility of these field descriptions, but refined them appreciably. Thus clay predominated in samples described as 'silty clay' in the field: in such cases the clay constituted more than 50 per cent by weight of the sample. In samples described as 'clayey silt' the clay and silt fractions were either about equal or silt predominated. 'Fine sandy' beds, however, were found to contain more silt particles than fine-grained sand. Within the predominantly clayey beds, lenses and partings of fine sand and silt were a common feature and in such beds the analysis of an individual sample which contained a particularly thick lens of fine-grained sand or silt commonly produced a result unrepresentative of the bed as a whole.

In general terms the basal 16 m of the succession is relatively arenaceous and laboratory tests prove that it contains up to 30 percent of fine-grained sand. The middle part of the sequence is more argillaceous, and the upper part shows a return of more arenaceous beds. Table 2 shows the six broad lithological subdivisions recognised in the Hadleigh Borehole; the limits of several of these units, however, are only approximate because of gradational contacts. In general, none of the units can be positively recognised by lithological criteria in an isolated exposure or in a single sample.

Fourteen diagnostic, facies-controlled faunal assemblages, which have been recognised in the Stock Borehole by King, Hughes and Wood (in Bristow, 1985), have been traced in the present district. These have been designated Units P1-14 in Figure 8 which illustrates the relationship between them, the lithological units mentioned above and the sub-divisions of Wrigley (1924), based on macrofaunal assemblages. Using all available evidence, a generalised outcrop map (Figure 9) has been constructed to show the likely broad distribution of the *lithological* units of the London Clay and the Claygate Beds, both at surface and beneath drift. In a recent paper King (1981) has subdivided the London Clay and Claygate Beds into five units (A to E of Figure 8). In the equivalent beds of the Hampshire Basin each of these units typically comprises a single coarsening upwards cycle. In Essex these cycles are not recognisable with certainty, and King (1981, pp.45–48) has used both lithological and faunal criteria to identify their correlatives.

Table 2 Lithological units within the London Clay recognised in the Hadleigh Borehole, together with the results of selected grading tests

Unit	Depth in Hadleigh Borehole (m)	Thickness (m)	Lithology	Selected test results of spot samples giving proportions of sand/silt/clay respectively
LF	26.99 to 47.09	20.10	Sandy silty clays and silty clays with abundant fine-grained sand laminae and partings	3, 37, 60; 6, 39, 55%
LE	47.09 to 70.40	23.31	Silty clays and fine-grained sandy silts with abundant fine-grained sandy lenses and partings. Glauconite is common in the lowest 4.5 m of the unit comprising up to 2 per cent by volume of the more sandy beds	7, 48, 45; 4, 36, 60%
LD	70.40 to 87.88	17.48	Clayey silts and silty clays with abundant fine-grained sandy lenses and partings	5, 43, 52; 5, 36, 59%
LC	87.88 to 130.76	42.88	Homogeneous silty clays and very silty clays with common large foraminifera and very few silty streaks	2, 37, 60%
LB	130.76 to 154.92	24.16	Clayey fine sandy silts with frequent silt streaks. This unit includes a 5 m silty clay bed	10, 40, 50; 5, 50, 45%
LA	154.92 to 159.19	4.27	Fine sandy silts with shells and scattered glauconite pellets and two calcareous cementstone horizons near the base	15, 55, 30%

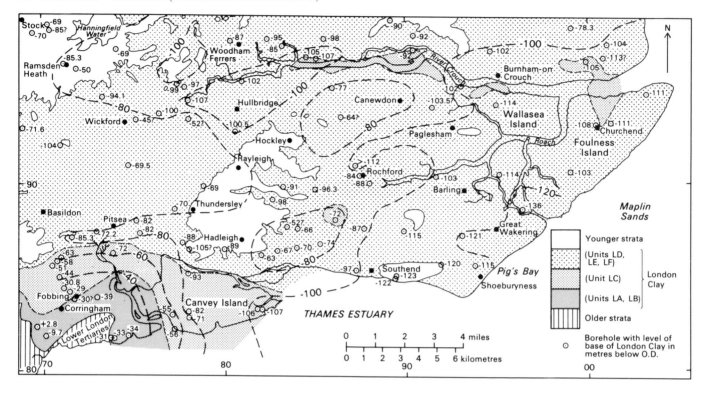

Figure 9   Contours on the base of the London Clay (20 m interval) and distribution of London Clay subdivisions and Claygate Beds

UNIT LA   Grain-size analyses of random samples within this unit generally show 15 to 30 per cent of fine-grained sand and up to 50 per cent of silt. In the Hadleigh Borehole, this unit comprised 4.27 m of dark grey silty fine-grained sand and sandy silt with a characteristic faunal assemblage (Unit P1 in Figure 8). Similar beds were recorded from the basal part of the London Clay in the Stock Borehole and, more recently, in BGS boreholes in the Braintree and Epping districts. In the well records of this district these sandy beds are either not differentiated from the London Clay or are included within the Oldhaven Beds.

Calcareous nodules are widespread in the basal beds of the London Clay. A septarian nodule (E44112*) between 159.04 and 159.10 m and a light grey calcisiltstone (E44111) from 158.25 to 158.54 m were both recovered from the Hadleigh Borehole, and a siltstone and sandstone were recorded from 3 m and 1 m respectively above the bottom of Stock Borehole. Prestwich (1850, pp. 253–254) notes 'calcareous semi-concretionary tabular masses' in the basal London Clay around Stifford Clays, Orsett and Stanford le Hope. The occurrence of calcareous nodules is discussed in more detail below (p.16).

Recent investigations at Harwich and Shotley, about 45 km north of Hadleigh (Knox and Ellison, 1979), have shown that the Harwich Member, which comprises the lowest 20 m of London Clay, correlated with Unit LA and

part of Unit LB, there contains numerous ash bands which correspond to a widespread volcanic 'Ash Marker' at the base of the Eocene in the North Sea. No conclusive evidence of tuffaceous material was recorded from Hadleigh, but Mr R. K. Harrison reports that specimens E44111 and E44112 (noted above) from Unit LA contain particles of suspected pyroclastic origin, although these are sparse and highly altered.

UNIT LB   Some relatively sandy beds and many very silty beds characterise this unit; up to 10 per cent fine-grained sand and up to 77 per cent silt were proved in grain-size analyses. However, there is a considerable degree of lithological variation, and silty clay beds with only 36 per cent silt are present. Some of the 'sand' and coarse-grained 'silt' noted in hand specimens have been subsequently shown to be gypsum crystals (see p.16).

UNIT LC   The beds within this unit are generally most uniform in lithology. They comprise compacted, commonly hard, brown-grey, silty clays with brittle fracture, interbedded with olive-grey, excessively fissured, silty clays which crumble on handling even when damp. A few clayey silt beds occur, generally towards the top and base of the unit, for example at 92.67 m depth in Hadleigh Borehole (4.79 m below the top of Unit LC), where there was an abrupt downwards change from olive-grey, stiff, very clayey silt to a compact, yellow brown, very silty clay. Both the brittle and the crumbly clay beds contain only about 2 per cent sand and around 60 per cent clay. Buff and light brown siltstone

---

* E Numbers refer to the English Sliced Rock Collection of the Geological Survey.

nodules occur at 114.64 m, 122.73 m and 127.56 m in the borehole (26.58, 34.85 and 39.68 m below the top of Unit LC respectively). These have darker brown cores and may be phosphatic.

Wrigley (1924, p.251) utilised the common occurrence of the crinoid *Balanocrinus subbasaltiformis* (Miller) within a restricted stratigraphical range as a criterion for establishing his London Clay 'Division Two' (see Figure 8). This marker fossil was found at 122.74 m and 123.01 m in the Hadleigh Borehole and also at a similar stratigraphical position in the Stock Borehole, thus establishing a provisional correlation with the London Clay sections zoned by Wrigley.

UNIT LD   Compared to Unit LC, there is an overall increase in silt content, which may rise to as much as 48 per cent, and the fine-grained sand fraction constitutes up to 10 per cent of the total weight as compared with only 2 per cent in the underlying unit. There is correspondingly a lower proportion of clay in these sediments.

UNIT LE   The presence of glauconite characterises many of the beds in this unit. It occurs as greenish brown or black, fine- to medium-grained, sand-grade pellets; the weathered clay contains abundant glauconite and has a distinctive, slightly speckled, 'pepper and salt' appearance. A few more sandy beds within this unit contain up to 8 per cent of fine-grained sand, and the silt content throughout varies between 35 and 50 per cent. The brachiopod *Terebratulina wardenensis* (Elliott) was found in this unit both in the Hadleigh and the Stock boreholes; it was also found in Unit LC in the latter.

UNIT LF   Beds within this unit are appreciably more sandy than those in units LB, LC and LD. The average sand content throughout is about 6 per cent, although thin beds and lenses of buff and dark green fine-grained sands up to 7 cm thick are common. Burrows are less common in this unit than in the lower ones and, as a result, primary laminations appear to be better preserved.   RAE

Outline of mineralogy

Mr R. J. Merriman of the Petrographical Department studied samples of London Clay, taken at approximately one metre intervals from the Hadleigh Borehole core (see Figure 10). He reports that micas, smectite and kaolinite constitute most of the clay minerals present. The micas comprise mainly detrital muscovite and clay mica, which is probably illite.

Muscovite is common, being most abundant and of coarser grain-size in the more arenaceous beds, particularly in Units LE and LF, where it is visible in hand specimens. Characteristically, flaky minerals such as muscovite are aligned parallel to the bedding planes (e.g. E 44339, E 44342, E 44343, E 44344, E 44345) except where bioturbation has disturbed the primary lamination (e.g. E 44340, E44341). Clay mica is most abundant in Unit LC, where it is accompanied by an interstratified mica/smectite mixed-layer mineral.

The content of kaolinite increases upwards in the borehole, varying from less than 10 per cent in Unit LA to

approximately 15 per cent in Unit LF. Chlorite, feldspars, apatite (collophane) and biotite are also noted in small amounts in the London Clay. Smectite, probably calcium montmorillonite, comprises up to 30 per cent of the London Clay and varies from below 20 per cent in arenaceous beds within Unit F to between 25 and 30 per cent in units LB, LC and LD.

Examination of thin sections, X-ray diffractometer traces, and subsequent calcimeter determination have proved that authigenic dolomite averaging 4 per cent by weight occurs throughout the London Clay down to a depth of 142 m (i.e. with the exception of only the basal 17 m), in the Hadleigh Borehole. Calcium carbonate (calcite) is restricted to Unit LC and is less abundant than dolomite, although minor amounts of calcite (less than 1 per cent) preserved in microfossils may not have been detected. In weathered London Clay dolomite has probably been selectively removed during weathering, which may explain why dolomite has previously been detected in only small amounts in the London Clay.

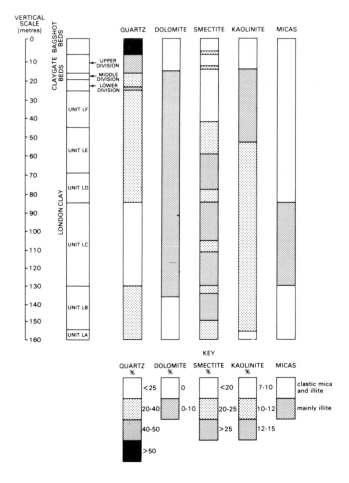

Figure 10 Generalised diagrams showing the mineralogy of whole rock samples from the London Clay, Claygate Beds and Bagshot Beds in the Hadleigh Borehole

Between 15 and 35 per cent by weight of quartz is recorded in the samples, decreasing from a maximum at the bottom of the London Clay to minimum values in Unit LC, and then increasing again through Unit LD to values of 30 per cent or more within the more silty and sandy beds of Units LE and LF.

Pyrite occurs throughout the London Clay, ranging from minute cubes and microconcretions visible only in thin sections to irregular-shaped nodules up to 3 cm in diameter.

Selenite (gypsum) crystals occur typically to a depth of about 3 m below the weathered zone of the London Clay. The selenite is generated by the interaction of sulphate ions, produced by oxidation of pyrite during weathering, and calcium carbonate present as shelly material in the clay. The 'weathered zone' of the London Clay normally refers to the completely oxidised brown clay, although, as selenite crystals are a product of chemical weathering by percolating groundwater, strictly speaking, the zone should be extended to include all the clays containing selenite. Small amounts of gypsum found in the lowest 17 m of London Clay in the borehole may have crystallised on the core since drilling.

Calcareous concretions, variously known as septarian nodules, cementstones or claystones, which are up to at least 40 cm in diameter and which characteristically show calcite veining, occur at irregular intervals throughout the London Clay. The septaria occur in beds and are not randomly scattered. Calcareous claystones may also occur without calcite veining. The precise origin of the septaria is not known, although bioturbation within some of the nodules indicates post-depositional lithification. Mollusc shells are commonly well preserved in nodules, whereas thin-shelled species (e.g. *Pholadomya*) are slightly crushed by compaction in the surrounding clays, showing that lithification occurred before the compaction of the sediments. In other instances, polyzoa and echinoderms are distributed solely on the surface of the nodules, suggesting that the septaria may have been 'solid objects resting on the sea bed' (Wrigley, 1924, p.248). Groundwater is often retained around nodules; in these circumstances excess pore pressures in the clay surrounding the nodules could result in mass-movement during excavations when the confining pressure of clay surrounding the nodules is released.

Glauconite pellets of fine- and medium-sand grade are found in the more sandy beds, between depths of 63 and 71 m in the borehole (i.e. at the top of Unit LD and the lower part of LE) and within the basal 4.5 m of the formation. Microscropic examination of sieved samples reveals the occurrence of small rounded and subrounded microcrystalline grains of dark green glauconite up to 0.12 m diameter throughout the London Clay.

RAE, GWG

Palaeontology

A stratigraphical succession of faunally based units has now been established for the London Clay of Essex. In many cases, it is the assemblage, rather than the ranges of individual species, which delineate the units. Units which are based on species ranges are biostratigraphical but may also be facies-controlled: units based on faunal assemblages are invariably facies-controlled. A correlation by means of these units is, therefore, not necessarily a chronostratigraphical one, although in the restricted area under review the time discrepancy is probably not significant.

The diagnostic facies-controlled assemblages that have been identified are shown in Figure 8. The majority of the fauna is benthonic and is, therefore, intimately associated with the sedimentary environment. Certain exceptions exist however. The planktonic foraminifera recorded in Units P8 and P9 are independent of the bottom facies and, because of their relatively rapid areal dispersion, can be regarded as representing a restricted time interval. They are thus valuable in long distance correlation. Because they are not facies-controlled, the pteropods which occur at the top of the London Clay and in the Claygate Beds are similarly useful for correlation in a part of the sequence which contains many rapid facies changes. A third group of planktonic organisms, the diatoms, which are present in the lower units of the London Clay are also good correlation indicators.

The London Clay faunal assemblages commence with the shallow water nearshore forms of Unit P1, which are followed by a predominantly arenaceous foraminiferal fauna that characterises Unit P2 and Unit P3/4. This is succeeded by open marine indicators with rich and varied assemblages of calcareous microfaunas but with only sparse macrofauna; these extend from Unit P5/6 to Unit P9. Above this, molluscs become increasingly important and the microfauna, although remaining common, is much more restricted in the number of species present. The lack of strong faunal differences between the units in this part of the succession (Unit P10 to Unit P14) commonly leads to difficulties in identifying the units from individual faunal samples. The topmost units (Unit P13 to Unit P14) reflect a return to shallow water conditions with rapid facies changes and the presence of the planktonic molluscs (pteropods).

MJH

Details

The London Clay has been proved in shallow boreholes and temporary exposures so numerous that it is not possible to list them individually. Similarly, calcisiltstone occurrences are not listed below unless their position in the sequence is known. More comprehensive information, together with details of the faunas, have been retained by BGS on file at Keyworth.

A well [7119 9648] at Tippler's Bridge, Ramsden Heath, proved 133.5 m of London Clay beneath Claygate Beds (based on Whitaker and Thresh, 1916, p.244). By comparison it is evident that the log of the 'Old Well' at Ramsden Bellhouse [7180 9519] must be in error: it recorded only 94.49 m of London Clay (Whitaker and Thresh, 1916, p.44), and it is now known that the well started in the lower part of the Claygate Beds. Indeed, many of the logs based, as is this one, on 'information from Mr. Purkis' are very generalised and commonly anomalous. The log of a well [7698 9647] at Old Rettendon Hall, also sunk by Mr. Purkis, is also suspect. This well which started about 4 m above the base of the Claygate Beds, is said to have proved 104 m of London Clay (Whitaker and Thresh, 1916, p.247).

A well [7743 9537] near the Rettendon Crossroads proved 126.18 m of London Clay (Whitaker and Thresh, 1916, p.246). The base of the Claygate Beds, which crops out some 30 m to the north-east is an estimated 3 m higher, giving a total thickness of about 129 m for the formation. A similar figure is obtained from interpretation of a well at Battlesbridge Railway Station [7792 9507]: the surface level of this well was given incorrectly by Whitaker and Thresh (1916, p.247) as 6.7 m above O.D. but is about 14 m above OD.

On the north side of the River Crouch, the foreshore is covered with cementstone nodules at the foot of 'The Cliff' [921 968]. It is calculated that the London Clay at outcrop in this vicinity is some 26 m below the base of the Claygate Beds and that the strata fall within Unit P13. A rich fauna and flora have been obtained from this locality (Kirby, 1974): although collected loose from the foreshore these can have originated only from this cliff. Of particular interest is the record of *Camptoceratops prisca* (Godwin-Austen) which has been recorded previously only in the Claygate Beds of Essex.

A well [7161 8718] at Vange Hall apparently commenced at about 18 m below the base of the Claygate Beds and penetrated 120.4 m of London Clay (Whitaker and Thresh, 1916, pp. 285–286). This comparatively large recorded thickness probably includes the basal part of the Claygate Beds.

The Salvation Army No. 1 Well [8024 8674] near Hadleigh proved 151 m of undifferentiated 'London Clay' (Whitaker and Thresh, 1916, pp.177–178). Comparisons with the nearby Hadleigh Borehole (see p.12) indicates that the uppermost 23 m recorded should probably be classified as Claygate Beds and Bagshot Beds.

In the Eastwood area, the Southend Water Works No. 4 Well [8277 8917], (Whitaker and Thresh, 1916, p.150), commenced 3 to 5 m below the base of the Claygate Beds and penetrated 132 m of London Clay beneath 4.3 m of Head (recorded as sandy loam and clay).   CRB, RAE

Trial holes and excavations in connection with the Prittle Brook Diversion Tunnel examined at the time of the survey showed that in the vicinity of the outfall at Chalkwell [8535 8555] the junction of the weathered and unweathered London Clay was very sharp, and extended inland as an almost horizontal surface, the weathered London Clay above it increasing in thickness from about 1 m to nearly 10 m. Traced seaward the unweathered zone maintains a depth of about 1 m below the beach deposits, which are themselves banked against a slope of London Clay at their inner edge. It thus appears that the cutting-back of the London Clay by the sea in Holocene times was a faster process than the development of the weathering profile in the clay, but that limited oxidation occurs underneath the beach deposits—at least along their landward edge.
                                                                    GWG

Claygate Beds

The Claygate Beds crop out extensively in the South Hanningfield and South Woodham Ferrers area, and around Rayleigh, Thundersley and Hadleigh. Smaller outliers occur at Westley Heights and Vange in the west, and at Althorne in the north-east. Palaeontological evidence from nearby shallow boreholes taken in conjunction with calculated depths to the base of the London Clay indicates that outliers of Claygate Beds probably crop out beneath drift deposits near Paglesham [9634 9274] and Beauchamps [9069 8837]; Figure 9 shows the location of the outcrops of the Claygate Beds and of the postulated subcrop below drift deposits.

The Claygate Beds comprise a broadly upward-coarsening sequence of interstratified silty clays, silts and fine-grained sands, which may be bioturbated. They range in thickness from 17 to 23 m. Their base is gradational but their junction with the overlying Bagshot Beds is typically sharp. The criteria used to define the Claygate Beds and their subdivisions are those put forward by Bristow, Ellison and Wood (1980).

The characteristic Claygate Beds lithologies observed from augering are buff sandy clay, typically rather uniform in colour and texture, and yellow and pale grey very sticky silty clays. The sandy clay may be confused with sandy wash derived from the overlying Bagshot Beds, whereas the silty clays may be difficult to distinguish from some beds in the upper part of the London Clay. A further source of difficulty in delimiting the Claygate Beds in the field is that, because of their clay content, drillers rarely distinguish them from the underlying London Clay. It is, accordingly, possible that the boundaries of the top and bottom of the Claygate Beds shown on the 1:50 000 geological sheet may locally have been drawn at rather too low a stratigraphical level.

Lithological divisions

The BGS boreholes at Stock, Hadleigh, Hockley and Westley Heights were used to formulate a standard succession for the Claygate Beds. A synthesis of further lithological and palaeontological information from field mapping, shell and auger boreholes, pits and sections has shown that three broad lithological subdivisions can be recognised in this district (see Table 3; Figures 11 and 12).

During the present survey the base of the Claygate Beds was taken at the lowest sandy horizon recognisable in the field. This is a change in practice for previous authors have regarded only those beds now assigned to the upper division as being the Claygate or 'Passage' Beds (Wooldridge 1924; Wooldridge and Berdinner, 1925). Thus, in the Hockley area, the total thickness of Claygate Beds as now defined is 17 m greater than that recorded by Wooldridge. Similarly, the Claygate Beds of other areas are increased by a thickness corresponding to that of the combined middle and lower divisions (Bristow, Ellison and Wood, 1980).

Characteristically the clays within the Claygate Beds weather from olive-grey (similar to London Clay) to a distinctive lilac colour, and then to an ochreous yellow-brown. The silts and sands vary in thickness but consistently occur at the same stratigraphical levels in the Claygate Beds (Figure 11). They weather to various hues of yellow and brown, varying with the concentration of secondary iron oxide. Springs are locally present where thick sand units occur near the base, and these give rise to a distinctive break of slope at the foot of scarp-like features. Head deposits mainly derived from the upper sandy beds and the Bagshot Beds may, however, totally or partially obscure the break of slope; in such cases seepages occur downslope from the Claygate Beds–London Clay boundary

LOWER DIVISION   Around Westley Heights, this division has a micaceous, glauconitic, bioturbated sand, containing about 45 per cent of fine-grained sand and 32 per cent of silt, at the base; it has a maximum thickness of 2.75 m. This is overlain by a homogeneous, silty clay bed, identical to strata in the upper part of the London Clay, which ranges in thickness from 2 m at Westley Heights to 5 m at Hadleigh. A

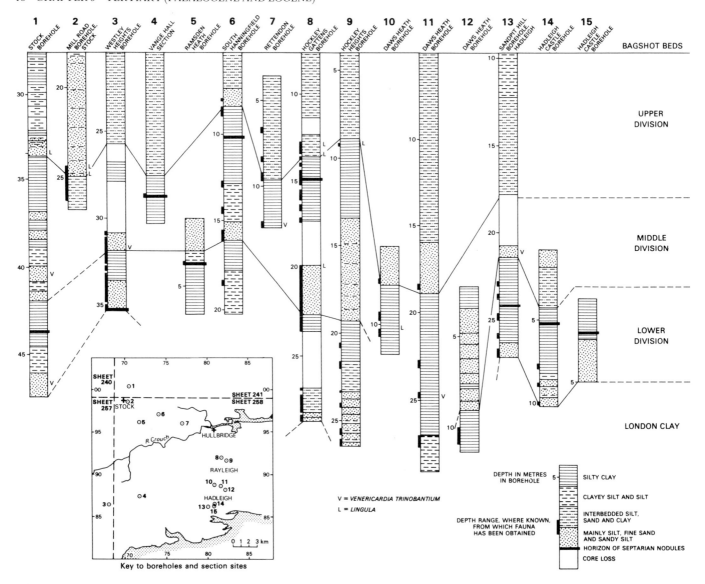

Figure 11 Correlation of the Claygate Beds in south-east Essex

septarian nodule horizon apparently occurs throughout the district in the middle of the clay unit.

MIDDLE DIVISION   The sandy beds within this division are thickest around Hockley and Stock where they are 4 to 5 m thick. The bivalve *Venericardia trinobantium* Wrigley is commonly present within these sandy beds (see Figure 11). The upper, more argillaceous beds are consistently 4 to 5 m thick, and commonly contain a septarian nodule horizon.

UPPER DIVISION   This part of the succession most closely resembles the Claygate Beds of the type area, and is well exposed at Vange Hall [7172 8758] where a thickness of 9 m was recorded. The section showed bioturbated silt and fine-grained sand, alternating with cross-bedded, finely laminated, micaceous sand, and containing pale grey clay flasers (Reineck and Wunderlich, 1968) many of which included small bioturbated sandy patches. Glauconite was present as green to black pellets giving a greenish tinge to the

weathered material, especially when it was wet. These sediments were probably deposited in a zone of fluctuating slack water and water turbulence, in a shallow-water marine environment where the water depth was possibly between 20 and 40 m (Reineck, 1964, p.192). A full description of this section is given in the details on p.20. Recent borehole investigations have proved similar strata up to 5 m thick near Daws Heath [807 882].

Towards the north and west of the district, the upper division becomes more silty in character and consists mainly of interbedded silty sands and clayey silts; in the Stock Borehole it is represented by 0.7 m of micaceous fine-grained sands, containing abundant *Lingula*, and an overlying 5.2 m of silts with thin sandy partings. In the Hadleigh Borehole 2.7 m of silts overlie 5.5 m of interbedded silts, micaceous fine-grained sands and clays; here, some of the fine-grained sand beds are similarly cross-bedded to those at Vange Hall, and thin iron pans and iron concentrations have developed within silts and sands in the highest 5 m.

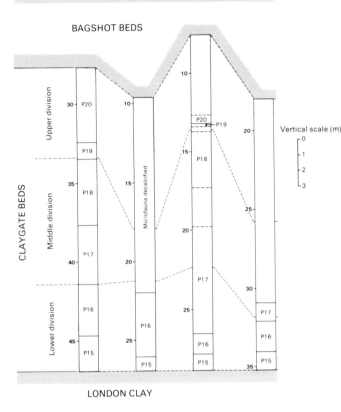

Depths recorded in metres from the surface in each borehole

Figure 12 Comparison of the lithological and palaeontological subdivisions of the Claygate Beds in cored boreholes

Outline of mineralogy

The petrographical characteristics of the Claygate Beds are generally similar to those of the London Clay (see Figure 10). Unweathered clays within the Claygate Beds are olive-grey in colour, but the more sandy beds are normally a dark greenish grey. Because of the more sandy nature of the Claygate Beds compared to the London Clay, groundwater percolates to a greater depth, thus increasing the depth of oxidation and selenite formation.

Mr Merriman records that the smectite content, which is generally between 15 and 20 per cent, is lower than that of the London Clay. Up to 15 per cent of kaolinite occurs in the lowest two divisions of the Claygate Beds, decreasing to 7 per cent in the upper division. Detrital muscovite is common, particularly in the more arenaceous beds. In the Hadleigh Borehole up to 5 per cent of dolomite is present in all but the upper 6 m of the Claygate Beds, and the quartz content fluctuates between 19 and 50 per cent by weight of the whole rock, reflecting the variability of the proportion of sand and silt grade material.

Fauna and environment of deposition

Detailed palaeontological work on unweathered Claygate Beds from borehole samples has been carried out by the Biostratigraphy Research Group of the Geological Survey and Mr. King of Paleoservices (see Appendix 2). The preservation of fossiliferous sections in the Claygate Beds is rare because of weathering. In the Stock and Hockley boreholes, where the Claygate Beds are relatively unweathered, six palaeontological units have been recognised (see Figure 12). These have been designated P15 to 20, so that they follow on, sequentially, from the faunal units of the London Clay (p.16).

A molluscan fauna similar to that in lithological Unit LF of the London Clay is generally characteristic of the Claygate Beds; one fossil unique to the Claygate Beds is the bivalve *Venericardia trinobantium*. Pteropods, being planktonic, are particularly useful for correlation and the pteropod *Camp-*

	Hadleigh Borehole	Stock Borehole	Lithology	Selected test results giving grain size proportions of sand/silt/clay (Hadleigh Borehole)
	Thickness (to the nearest 0.5 m)			
Upper division	8.5	6.0	Interbedded fine-grained sands and silts, with subordinate clay lenses and streaks	32, 60, 8%
Middle division	3.0	6.5	Silty clays and clayey silts with micaceous fine-grained sand laminae overlying finely laminated, micaceous, glauconitic silty fine-grained sands	37, 50, 13; 23, 62, 15%
Lower division	6.0	5.5	Homogeneous silty clay overlying micaceous, glauconitic, bioturbated fine sand or clayey sand	4, 46, 50; 10, 49, 41; 43, 27, 30%

Table 3 Lithological divisions within the Claygate Beds recognised in the Hadleigh and Stock boreholes, together with the results of selected grading tests

toceratops prisca is found principally in the Claygate Beds (Bristow and others, 1980), but has been recorded by Kirby (1974, p.10) from within lithological Unit LF of the London Clay at The Cliff, west of Burnham-on-Crouch. *Lingula* is commonly found in the upper Claygate Beds (see Figure 11) and, by virtue of its chitinous shell, may also be preserved in decalcified beds.

The fauna of the Claygate Beds includes common marine browsing, suspension-feeding, and carnivorous molluscs. The high proportion of the two former types suggests an environment in the photic zone, but no forms typical of very shallow water are present. The high numbers of pteropods at certain horizons suggest slow or interrupted deposition, leading to the concentration of fossils.   R.AE

Details

Logs of selected borehole sections together with brief notes on the faunas obtained from the boreholes appear in Appendix 2.

Stock-Bushy Hill area

Immediately south of Stock generally only yellowish brown silty clay was proved by augering. The upper junction with the Bagshot Beds is sharp and marked by a spring line in many places.

Fine-grained sands at or near the base of the Claygate Beds have been observed in the following localities south of Stock: [6875 9394; 6879 9712; 6883 9755; 6918 9448; 6946 9378; 6948 9338; 6961 9343].

Althorne area

The presence of Claygate Beds in this vicinity has not been recognised previously. They do not form a strong feature, although they cap high ground. Auger evidence indicates the presence of yellowish brown silty clays; interbedded sands have rarely been noted. Where the tongue of Claygate Beds extends south-eastwards from Althorne towards Burnham-on-Crouch the base of the formation is mappable only with difficulty, and its boundary with the London Clay is conjectural.   CRB

One Tree Hill-Vange area

Dines (in MS. 1923) described a former brick-pit section at One Tree Hill [6962 8607] as:

	Thickness m
Loamy soil with pebbles	0.6
Yellow sand with some clay seams	
Small faults	2.1 to 2.4
Lilac clay and sand, laminated in approximately	
25 mm seams	—

In 1972 the exposed section showed:

HEAD
Pebbly wash and sandy loam	1.5

BAGSHOT BEDS
Sands, finely laminated, buff with lilac-brown clay partings up to 8 mm thick	1.4

CLAYGATE BEDS
Clays, lilac-brown, with ochreous silty partings; bioturbated	2.3

RDL

The beds exposed at Vange Hall Brickpit [7172 8738] were briefly described by Wooldridge and Berdinner (1925, p.322). Dines (in MS. 1923) noted the following section:

	Thickness m
Earth loam with pebbles	0.9
'Bagshot Beds' in a small synclinal depression	0.6
Laminated clays and sands with sand beds up to 0.3 m thick	7.5 to 8.5

In 1974, parts of the degraded face were cleared and the following section was exposed:

	Thickness m	Depth m
Sand, fine-grained, silty, brown, micaceous	0.4	0.4
Interbedded sands, silts and clays; bioturbated dark greyish brown clay predominates, with silty sand partings at about 3 cm intervals	1.1	1.5
Sand, fine-grained, finely laminated	0.3	1.8
Silt, clayey, bioturbated	0.2	2.0
Sand, silty ?(section obscured)	1.4	3.4
Silt, clayey, bioturbated, greyish brown	0.4	3.8
Sand, fine-grained, with sandy silts and silty clay flasers	0.9	4.7
Sand, silty ? (section obscured)	1.6	6.3
Interbedded sands, silts and clays	1.2	7.5
Silt, clayey, with local sandy intercalations, bioturbated; glauconite pellets	1.3	8.8
Section obscured	1.0	9.8
Clay, silty, greenish brown with abundant sandy streaks and partings; a septarian nodule lies 0.5 m from the top of the exposed bed	2.2	12.0

The upper part of the section to 8.8 m depth, is assigned to the upper division of the Claygate Beds and the lowest beds exposed are referred to the middle division. Septaria collected from the lowest silty clay bed (Wrigley in MS.) yielded the following mollusca: *Cultellus affinis* (J. Sowerby), (common), '*Striarca*' *wrigleyi* (Curry), *Corbula globosa* J. Sowerby, *Ficus multiformis* Wrigley, *Euthriofusus coniferus* (J. Sowerby) and *Tibia lucida* (J. Sowerby).

South Benfleet area

Near Round Hill, small exposures are present behind the active landslips to the east of the Bagshot Beds outlier; these sections show ochreous clayey silts with sandy partings.

Hockley-Rayleigh-Hadleigh area

An old brickworks at Down Hall [810 918] has been backfilled and no section was visible in the Claygate Beds during the survey. Wooldridge and Berdinner (1925, p.114), however, gave the following succession: Bagshot Sands 14 ft (4.3 m); Passage Beds (= Claygate Beds) 14 ft (4.3 m); London Clay 15 ft 4 in (4.7 m). Mapping of this ground (by MRH) suggests that the bed recorded as London Clay is a clay within the Claygate Beds as defined in this memoir (see Figure 11). A fauna from this locality has been recently re-examined by the BGS Biostratigraphy Research Group (Bristow and others, 1980) and otoliths from the same locality have been described by Stinton (in Casier, 1966).

BGS boreholes at Hockley Gattens [8176 9205] and Hockley

Heights [8244 9158] penetrated 23.3 m and 22.5 m of Claygate Beds respectively, beneath Bagshot Beds (see Figure 11). The three divisions of the Claygate Beds were recognised in both holes, although the basal sand of the Claygate Beds was not penetrated in the latter one. In the Hockley Gattens Borehole, *Lingula* was common between 19.74 m and 20.82 m, and from 13.19 m to 13.40 m (upper division): in the Hockley Heights Borehole, *Lingula* was noted in the upper part of the middle division at a depth of 9.2 to 9.4 m.

A number of trial pits near West Wood, Daws Heath [808 882] exposed the upper division and the upper clay of the middle division of the Claygate Beds. The upper division consists mainly of silty fine-grained sand, finely laminated in places, with beds containing greenish and pinkish grey clay laminae.

Temporary excavations near Rayleigh Weir proved the existence of a comparatively thick sand in the upper part of the Claygate Beds.   CRB, RAE, GWG, RDL

South of Hadleigh, the base of the Claygate Beds is mostly obscured by slips, but boreholes show it to range between 43.6 m OD in the Hadleigh Borehole and 47.3 m OD in a borehole at Castle Farm, Hadleigh [8086 8646] (Appendix 2).   RAE

Bagshot Beds

Most of the high ground in the present district (i.e. above 60 m OD) is capped by Bagshot Beds. The four main outliers (see Figure 1) are: West Hanningfield-Stock area; the One Tree Hill and Westley Heights area; the Hadleigh-Thundersley area and the Hockley Woods area.

Lithological divisions

In south Essex the top of the Bagshot Beds has been removed by post-Eocene erosion and hence their original thickness is not known. The thickest remaining sequence was proved in the Stock Borehole where 27.58 m of Bagshot Beds were recorded and the following three-fold sequence was established (Bristow, 1985): Pebble Bed, 4.22 m; clay and silt, 10.08 m; sand, 13.28 m. In the following account these three lithological divisions are referred to as the Sand division, the Clay-and-silt division and the Bagshot Pebble bed. On the 1:50 000 geological map, the Pebble Bed is termed the Bagshot Pebble Bed, and the lower two subdivisions are grouped as the Bagshot Beds.

About 1 km to the south of Stock, in the extreme north of the district, a BGS borehole [7020 9988] proved 15.2 m of Bagshot Beds as follows:

	Thickness m	Depth m
Sand, fine-grained, silty, with rounded brown flint pebbles	0.5	0.5
Interbedded sand, silt and clay	2.7	3.2
Sand, fine-grained	3.3	6.5
Sand, fine-grained, with beds of silt and clay	2.8	9.3
Sand, fine-grained	seen to 5.9	15.2

The highest (pebbly) bed may be a remnant of the Pebble Bed, and the underlying strata to a depth of 9.3 m are possibly comparable with the Clay-and-silt division in the Stock Borehole.

SAND DIVISION   This comprises uniform, well-sorted, finely laminated, slightly micaceous, fine-grained sands. At outcrop the sands weather to an orange-brown colour, with thin local iron-pans; springs are common at the base. Subordinate beds, up to 5 cm thick, of lilac and pale grey clay, commonly containing small black carbonaceous flecks, occur throughout. Grain-size analysis shows the sands to consist of up to 70 per cent of fine-grained sand with 30 per cent silt.

The maximum thickness proved in south-east Essex is 13.28 m in the Stock Borehole. The more southerly and easterly outliers in the present district show up to 7 m of sands on Daws Heath [805 885], up to 10.5 m at Sandpit Hill, Hadleigh [801 865], and up to 7 m around Rayleigh. A clay bed within the Sand division analysed by Mr Merriman contained 30 per cent of smectite compared with only 15 to 20 per cent in clays belonging to the Claygate Beds. Conversely, kaolinite constituted a lower proportion of the clay minerals than in the underlying beds.

The sands were deposited in a shallow sea, as evidenced by the presence of marine molluscs including *Ringicula sp.* and *Adeorbis sp.*, which were preserved in limonite in the Westley Heights Borehole. Salter (1907) recorded bivalve shell casts at Westley Heights, and Wooldridge and Berdinner (1922) mentioned the presence of ferruginous casts resembling *Pectunculus* and *Cyprina* in the Bagshot Beds at One Tree Hill. Wooldridge (1924, p.364) also recorded indeterminate mollusc casts and shark's teeth in the Bagshot Beds at Rayleigh, Hadleigh, Stock and elsewhere in south Essex. RAE

CLAY-AND-SILT DIVISION   This division is preserved only in the West Hanningfield-Stock area, where it reaches a maximum thickness of 10.08 m in the Stock Borehole. Whitaker (1872, p.327) referred to this division as 'brickearth'. The 1.2 m of 'yellow and blue clay' beneath 'coarse gravel' and overlying sands in a well [6917 9909] north of Stock High Street, may also be referable to the Clay-and-silt division.

East of Stock, a BGS borehole [7031 9853] proved 3.0 m of silty clays and fine-grained sandy silt, locally laminated, overlying the typical sands of the Bagshot Beds. The borehole was sited north-west of an old brickpit [7045 9845] in which H. G. Dines (in MS, 1923) noted 1.5 to 1.8 m of laminated sands and grey lilac clays 'similar to Claygate Beds'. Farther south, silty clays encountered in augering in the south-eastern part of Great Bishop's Wood [705 975] probably belong to this subdivision.   CRB, RAE

BAGSHOT PEBBLE BED   The pebble bed is preserved only in the Stock area. The only well documented exposure of it in the London Basin is at South Weald, near Brentwood, to the west of the district, where it comprises monogenetic black well-rounded flint pebbles in a fine-grained sand matrix; the pebbles show arcuate fracturing, either as a result of beach battering or pressure. Pebbles in the appropriate position in the Stock Borehole are well-rounded black flints up to 3 cm in diameter, and Wood (1868, p.464) described 'pebbles in the form of bands embedded in laminated clay' near Stock. Other pebbly occurrences, similar to those at Stock and South Weald but including vein-quartz and other exotic pebbles, have probably been derived by solifluction from glacial gravels and the Bagshot Pebble Bed.

Evidence from the Westley Heights area, west of the present district, indicates that 1 to 4 m of pebble bed generally overlie about 9 m of Bagshot Beds sand; at Stock about 23 m of Bagshot Beds underlie the pebble bed. The Westley Heights Borehole recovered black rounded flint pebbles with a maximum diameter of 15 cm, showing strong arcuate fractures. It was not possible to record the exact thickness of this bed, but at least 4 m of pebbles were penetrated. In a small section near 'The Crown', Westley Heights [6805 8671], Salter (1907) recorded a gravel composed of 'Tertiary flint pebbles, sub-angular flints, and sub-angular and rounded pieces of Lower Greensand Chert'; this suggests that the pebble bed at Westley Heights is not the Bagshot Pebble Bed. RAE

Details

Stock-Ramsden Heath area

In general there is a gentle southward dip to the Bagshot Beds in this area, as a consequence of which copious springs issue from the base of the formation to the south of Stock village [around 690 985, 693 984, 697 984, 700 979] and enable the contact with the Claygate Beds to be mapped with ease, although the boggy terrain and the presence of Head in such areas may locally obscure the boundary.

In a well [6917 9909] just north of Stock High Street, 8.38 m of sand were proved beneath 1.2 m of 'brickearth', and overlying the Claygate Beds. Another well [6917 9909] on the north side of the High Street penetrated: Topsoil, 0.91 m; yellow clay and pebbles, 3.05 m; yellow and blue clay, 0.46 m; coarse gravel, 0.30 m; overlying 9.6 m of brickearth and sand of the Bagshot Beds. The pebbly beds, presumed to be the Bagshot Pebble Bed, have been worked at a number of locations in the past [6903 9882 and 6930 9890].

A borehole [7031 9853] proved one of the greatest thicknesses of Bagshot Beds sands in the area, although the thickness may have been exaggerated by 'blowing sand' which was a serious problem in this borehole. Some 20 m of laminated fine-grained sand rested on the Claygate Beds, and was overlain by 3.8 m of 'brickearth'. CRB

Westley Heights area

In the Westley Heights area, the Bagshot Beds appear to consist predominantly of fine-grained sands. A locally extensive surface wash of pebbly loam may in part be derived from the Bagshot Pebble Bed.

Rayleigh-Thundersley area

At Hambro Hill, Rayleigh cross-bedded sands with sets of cross-strata up to 80 cm thick are exposed [8132 9191]. Convolute structures were noted in the sands at this locality.

Excavations for a swimming pool [7926 8800] near Kiln Road, Benfleet, proved:

	Thickness m
HEAD	
Ochreous pale grey silty clay, structureless	1.4
BAGSHOT BEDS	
Fine-grained sands, ochreous pale grey, rare bedding structures; pale grey clay parting 0.5 m from the top	1.7

Sands were reported to have been proved to a depth of 5.2 m below the recorded section. RDL, RAE

Over much of Thundersley-Benfleet outcrop, fine-grained sands were proved by augering. Locally clayey and silty beds were noted, particularly in the Great Common area. A pit in Common Approach, Thundersley [7922 8910] showed:

	Thickness m	Depth m
HEAD	1.75	1.75
BAGSHOT BEDS		
Clay, silty, brown, cryoturbated, with beds of greenish clay up to 10 cm thick and laminae of very fine-grained micaceous sand	0.18	1.93
Interbedded pale brown and green mottled silty clay, with clayey silt and clayey, fine-grained sand laminae up to 1 cm thick	0.42	2.35
Clay, very silty, finely laminated with lenses of light, yellow-grey, micaceous fine-grained sand up to 1 cm thick; sharp base	0.32	2.67
Sand, fine-grained, yellow-brown, micaceous, silty with clay laminae and rounded clay clasts representing reworked clay laminae (basal 1.3 m augered)	2.23	4.90

RAE, RDL

An exposure [7993 8675] in the pit complex at Sandpit Hill, Hadleigh, showed the following section:

	Thickness m	Depth m
MADE GROUND AND HEAD deposits	1.4	1.4
BAGSHOT BEDS		
Sands, fine-grained, and silts, ochreous, finely laminated; pale grey clay partings up to 0.03 m thick	3.0	4.4

RDL

Daws Heath area

Where the outcrop is relatively free of drift the base of the Bagshot Beds is usually marked by a feature which may be associated with small seepages. North of Daws Heath and between Hadleigh and Great Wood, however, sand loam and gravelly drift obscure the base of the Bagshot Beds and the mapped line is very indefinite.

A BGS trial pit [8076 8824] at West Wood, Daws Heath exposed:

	Thickness m	Depth m
HEAD		
Pebbly micaceous fine-grained sand and clayey sand	0.65	0.65
BAGSHOT BEDS		
Sand, micaceous, fine-grained, cross-bedded and finely laminated, yellowish orange, iron-stained in places; cryoturbation interrupts the primary structures with pipes of structureless fine-grained sand extending to 2.5 m depth	3.10	3.75
CLAYGATE BEDS	—	—

RAE

CHAPTER 4

Structure

The Southend district, in common with much of the Thames Estuary region, is one where the geological structure is comparatively simple. Although a few faults (generally with throws of less than 30 m) have been detected by detailed field mapping in the London area and south-east of Chelmsford, none has been proved in the present district by this method. Geophysical surveys have, however, suggested the presence of localised zones of small faults which may be partly associated with present-day major valleys (see below). Cored borehole samples, notably from the Hockley Borehole [8176 9205], show bedding-plane shears within the London Clay, and it may be inferred that appreciable deformational stresses have been absorbed by this mechanism within the Tertiary strata.

The structure contour map of the base of the Tertiary (Figure 4), and the outcrop patterns of the Claygate Beds and Bagshot Beds, show that the Tertiary and Mesozoic strata are folded into a number of broad, apparently symmetrical, flexures which trend approximately WSW–ENE. There is insufficient borehole evidence to indicate whether these structures are co-linear, with varying amplitudes along their length, or whether they are arranged in an *en échelon* pattern: the latter seems more probable.

To the south of the district at Cliffe, Kent, an eastward-trending anticline flexure, which is of comparable scale to the above and causes Upper Chalk to crop out, can be shown to be related to a graben in the underlying Palaeozoic, Jurassic and Lower Cretaceous strata (Owen, 1971). One or both of the faults which bound the graben moved in early Cretaceous times, in pre-Upper Gault times, and again in post-Eocene times. In the last period the fault-block moved up instead of, as formerly, down. This caused the post-Jurassic sediments to 'drape' over the graben in an anticlinal fashion. As the fractures do not penetrate the post-Jurassic rocks it is possible, particularly in view of the relatively small thickness (about 300 m) of strata involved, that the flexures in the Southend district are reflections of similar fault-bounded blocks in the underlying Palaeozoic basement.

Evidence from geophysical surveys (Shephard-Thorn, Lake and Atitullah 1972, p.100), indicates that the gravity anomalies in south Essex have a pronounced regional NW–SE trend which may be related to the distribution of pre-Carboniferous magnetic basement at relatively shallow depth (Terris and Bullerwell, 1965; Watson, 1980, pp.5–8), rather than any major structural grain within the younger Palaeozoic rocks.

Although apparently tectonically stable for a considerable period of geological time, the district has, by virtue of its proximity to the Southern North Sea Basin (Rhys, 1974), been influenced by the earth-movements there. More particularly, the continuing Quaternary downwarping of the basin has been responsible for progressive subsidence of the Thames Estuary region. Where Flandrian sediments are preserved, the effects of tectonic subsidence are complicated by factors such as the varying rates of sediment compaction. Embankment and drainage works have restricted the sediment input to the marshlands and thus disrupted the natural balance between sedimentation and compaction. Isostatic readjustments, subsequent to the last glaciation, and eustatic variations in sea-level have also affected the area. Because of the interplay of these various influences, the problems of resolving sea-level changes are complex. The overall trend of sea-level variation has been tentatively suggested to be a rise of from 1.0 to 1.3 m per century over the last seven to nine thousand years (see for example D'Olier 1972; Akeroyd 1972). Sea-level observations at Southend and Tilbury (Rossiter, 1972) give a calculated relative rise in sea-level of just over 3 mm per year over the last fifty years.

DETAILS

Mr M. Sarginson, formerly of the BGS Engineering Geology Unit, reports that marine seismic surveys on the River Crouch, to the west of the grid easting 93 in the vicinity of Black Point [913 967], show the presence of two reflecting horizons within the London Clay. These seismic markers, which may be calcisiltstone layers, show the presence of small-scale faulting, with a maximum throw of about 3 m, and of both small-scale and large-scale folding. The trend of the faulting shows a consistent NE-SW pattern, whereas a WNW–ESE trend is suggested for the folds. It is not clear whether these structures are tectonic or whether they represent valley-bulges. The River Crouch has a buried channel extending down to about 18 m below OD in this area, and the latter explanation may accord with this over-deepening.

The closed area of negative gravity anomaly centred in the Thames Estuary, south-east of Canvey Island was investigated by the Canvey Island Borehole [8215 8330] (Smart and others, 1964). The borehole proved Cretaceous strata to rest directly on Old Red Sandstone. Bullerwell (in Smart and others, 1964, p. 34) calculated that the gravity deficiency could be accounted for by a thickness of Old Red Sandstone of the order of 1000 m.   RDL.

CHAPTER 5

Quaternary: Pre-Hoxnian

Within the district the Tertiary rocks are partly concealed beneath Quaternary deposits (Figure 2). Figure 13 is a rockhead contour map of the eastern part of the district. It combines subdrift contours, where the eroded top lies below Quaternary deposits, with present topographical contours where the Tertiary rocks crop out at the surface.

As far as possible, the Quaternary deposits have been described below in chronological order, chapters 5, 6 and 7 describing Pre-Hoxnian, Hoxnian to Devensian, and Flandrian deposits respectively. However, it is difficult to separate buried Devensian and Flandrian sediments in some of the areas covered by Flandrian marine alluvium; accor-

dingly Chapter 7 includes a description of buried channels and associated deposits, some of which are known to be or pre-Flandrian age.

Deposits of undoubted glacial origin are confined to a few small patches on the higher ground in the north-west of the district; both Boulder Clay and Glacial Sand and Gravel are present. They form scattered outliers marginal to a much more extensive sheet of deposits which stretches northwards and north-eastwards into Suffolk. The fragmentary nature of the deposits in the Southend district is probably in part an original depositional feature and in part due to post-depositional erosion. In either event it hampers interpretation.

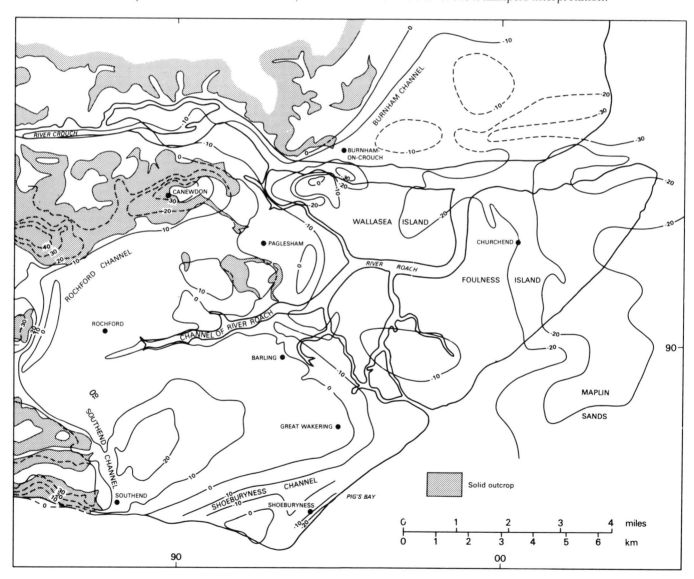

Figure 13 Contours, in metres relative to OD, on the eroded top of the London Clay or Claygate Beds

BOULDER CLAY

In East Anglia the Boulder Clay was called the 'Great Chalky Boulder Clay' by Harmer (1904; 1909). Baden-Powell (1948) sought to divide it into a lower 'Lowestoft Boulder Clay' (Lower Chalky Boulder Clay) and an upper 'Gipping Boulder Clay' (Upper Boulder Clay), which he considered were deposited during distinct glacial episodes separated by the Hoxnian inter-glacial. A study of the orientation of the long axis of included pebbles by West and Donner (1956) tended to support this view. Clayton (1957; 1960) suggested a local nomenclature for various litho-stratigraphical units which he recognised in this general area: in particular he recognised and named three distinct tills. He considered that all the patches of chalky Boulder Clay within the Southend district belonged to his 'Hanningfield Till', which he considered to be the oldest in the sequence. Later work has, however, revised this interpretation, and Bristow and Cox (1973) consider that there is no valid reason for regarding the 'Hanningfield Till' as distinct from the 'Springfield Till', thought by Clayton to be the youngest one in the region. They also believe that all the till in the Chelmsford district is what was formerly termed the Lowestoft Boulder Clay. The present survey supports this view. In northern Essex and Suffolk deposits of Hoxnian age overlie this till (Turner, 1970), which has been ascribed to the Anglian Stage (Mitchell and others, 1973), believed immediately to precede the Hoxnian inter-glacial.

The chalky Boulder Clay occurs as isolated masses around South Hanningfield. The till is thought to have resulted largely from the in-situ wasting of the ice-sheet. It is a medium bluish grey to greyish brown clay. A weathering zone, indicated by pale grey and yellow mottling, extends down to a depth of some 1.5 m. Where the till overlies impermeable material such as London Clay, it is a uniform medium grey down to its base: calcareous 'race' may, however, be present in the substrate. Where permeable deposits underlie the till, the chalk detritus has been dissolved to a varying extent and the base of the till is brown.

The Boulder Clay occurs mainly as outliers which are discrete from, and more widespread than, those of Glacial Sand and Gravel. Where the two lithologies are seen in juxtaposition, as on a hill [707 944] to the west of Ramsden Park, the sand and gravel overlies the Boulder Clay. However, because these sands are ice-marginal deposits possibly laid down on the flank of a melt-water overspill channel (Clayton 1957, p.11), this relationship is not considered temporally significant.   RDL

Details

Most of the outcrops are small and there are no permanent exposures. Where proved by augering, the Boulder Clay is characteristically chalky to within 0.5 to 1 m of the surface, containing abundant chalk pebbles, chalk-flour, a high proportion of flint pebbles and a variety of erratic pebbles.

Numerous pits were formerly opened in the chalky Boulder Clay outcrops, presumably to work the deposits for marl. The shallow depressions that remain form good indicators of the presence of till. Where the till is overlain by extensive sheets of solifluction material, as for example to the south-east of Stock, old pits which were dug through the Head to it now form shallow depressions. At Downham, temporary sections [around 7220 9643 and 7260 9641] for a gas pipe-line showed up to 1.5 m of mottled orange and grey sandy clay overlying chalky Boulder Clay. A somewhat thicker spread of similar material was noted at Frennels [722 971] by Clayton (1957, p.9), where up to 3 m of orange sandy and pebbly clay rests on some 2 m of chalky Boulder Clay. He interpreted this as the deeply weathered upper surface of an ancient till—the Hanningfield Till—and used this criterion to separate such exposures from those of the younger Springfield Till. It now seems more probable that the surface deposit is a Head, and accordingly the separate existence of the Hanningfield Till is thought to be unlikely. Similar sections [7120 9535; 7112 9523] exposed up to 1.8 m of clayey rounded flint gravel overlying the Boulder Clay; the junction between the two deposits in the latter section was almost vertical.   CRB

GLACIAL SAND AND GRAVEL

Most of the sparse deposits of Glacial Sand and Gravel in the district are thought to have been laid down extra-glacially by outwash streams draining the Anglian ice-sheet, for the distribution of glacial deposits in the adjoining Chelmsford district indicates that, in general, the ice-sheet terminated along the Danbury-Tiptree ridge, and incursions of ice into the present district appear to have been restricted. However, some of the sand and gravel may have been laid down by streams that flowed on the ice; this may explain the sands near Ilgar's Manor [787 990] and west of Ramsden Park [705 941]. Detailed information from these localities is limited, but the deposits seem comparable to those around Tiptree.

Details

There are a few significant exposures of sand and gravel in the district. Two patches of sand and gravel, to the south of Ramsden Heath have been worked in the past. In 1923 Mr F. H. Edmunds noted 1.8 m of sand and gravel with large nodular flints, vein quartz and Bunter sandstone pebbles in a pit [7073 9438] close to the hill crest, and noted 1.2 m of gravel in two other nearby pits [7073 9426 and 7050 9413]. A nearby borehole [7053 9400] near Ramsden Park proved 4.7 m of sand and gravel overlying Claygate Beds.   RDL, CRB

SAND AND GRAVEL OF UNKNOWN ORIGIN

Outcrops of sand and gravel near Ashingdon [855 933], Upper Parkgate [824 913], Rayleigh, Claydons [801 890], Daws Heath, and Hadleigh cannot be directly related to the glacial deposits nor to the river terraces. The surfaces of these spreads range in height from 55 to 80 m above OD, and in the two largest outcrops, at Daws Heath and Hadleigh, fall away gently to the east, suggestive of either a terrace or an outwash plain. Three shallow boreholes [8220 8821; 8160 8866; 8076 8883] which penetrated these deposits on Daws Heath proved thicknesses of sandy gravel ranging from 0.9 m to 5.5 m. These variable thicknesses, coupled with evidence obtained from adjacent field-mapping, show that the gravels have an irregular base possibly because of post-depositional cryoturbation.

These deposits predate the bulk of the terrace gravels. Their relatively great height suggests that they are of some antiquity, possibly dating back to before the Anglian glaciation, and they may have been laid down by south-bank tributaries of a proto-Thames that followed a course well north of the present river (Bridgland, 1980). Their grading characteristics, and in particular their high proportion of 'fines', show a close resemblance to undoubted glacial sands and gravels, and contrast with the local terrace deposits. Wooldridge (1923, p.320) named these gravels the Rayleigh Gravels, and described exposures in them which showed interbedded coarse-grained ferruginous sands and gravels with large-scale cross-bedding. The pebble content was noted to include Tertiary-derived flints, large slightly worn flint nodules, angular and rounded Lower Greensand cherts, Carstone pebbles and 'smaller pebbles chiefly of quartz and lydian stone'. He noted striations on the larger blocks and inferred from the shape of the flint nodules that transport by floating ice was the most likely agency for some of the gravel content.

Details

The gravels have been extensively worked in the past from shallow pits in the Daws Heath area. Mapping indicates that their base is slightly irregular, especially in the eastern half of the spread and also along their western edge where a borehole [8076 8883] has proved 5.5 m of sandy gravel within 125 m of the margin of the deposit, thus giving a minimum gradient for the base of 1 in 23. The stone content of the gravels almost exclusively comprises partly-worn to well-worn flints. A single sarsen 0.4 m^3 in volume, which was apparently left from adjacent gravel workings, lay on the surface [8088 8900], and a smaller one was seen amongst gravel debris farther east [8243 8856]. A trench section [8124 8875] near the centre of Daws Heath showed 2.1 m of yellow sand with thin gravel partings beneath 0.8 m of sandy top soil. Up to 2 m of similar sands were proved in a temporary exposure [8220 8836] near Oakwood Reservoir.

The only other noteworthy section is in London Road, Hadleigh, where an excavation [8185 8717] exposed 2.1 m of yellow medium to coarse-grained sand with gravel partings overlying sands of the Bagshot Beds. The base of a large lobe of sand and gravel, apparently at the southern extremity of the spread around Hadleigh, falls southwards west of Park Farm [813 865] through a vertical height of 13 to 14 m within a horizontal distance of about 340 m. This lobe appears to be the remnant of a local drift-filled channel, most of which has been removed by erosion.   RDL, GWG

CHAPTER 6

Quaternary: Hoxnian to Devensian

Hoxnian to Devensian deposits cover a considerably greater area within the district than the pre-Hoxnian deposits; in addition to those deposits which lie at the surface and, therefore appear prominently on the 1:50 000 map, they include the fill of an extensive system of buried channels. This fill is thought to be largely of Hoxnian age (see below). River deposits in the valley of the River Crouch and between Paglesham and Southend are related to a proto-Thames/Roach system; others around Stanford le Hope are related to the present Thames. Brickearth and Head Brickearth are also present, particularly in the Southend area, and Head is widely distributed throughout the district.

Certain of the drift deposits, particularly those in the Southend-Paglesham area, do not have a single simple origin. The River Terrace deposits, Head, Brickearth, Head Brickearth and buried channel deposits were originally laid down in a variety of environments, including estuarine, fluviatile and aeolian, but subsequent remobilisation by cryoturbation and solifluction, together with subaqueous redistribution, has led to the evolution of complex modified sequences. Each of the drift deposits is described below under the broad subheading that appears most appropriate on the basis of its genesis and lithology, though in some instances this involves gross simplification. The genetic problems are possibly most acute in the case of the brickearths. Certain of the Brickearths, Head Brickearths and river terrace loams have grading curves resembling those of typical loess (Gruhn, Bryan and Moss 1974; Perrin, Davies and Fysh 1974). However, although samples of comparable brickearth material from north Kent have been shown to have a mineralogy distinct from that of local bedrock (Weir, Catt and Madgett, 1971), suggesting an aeolian origin, this is apparently not so in the present district. Their origin is, therefore, not conclusively established, though it is probable that a significant proportion of loessic material has been incorporated in the loams and has been subsequently remobilised.

BURIED CHANNEL DEPOSITS (ONSHORE)

Deep drift-filled depressions have been proved in the Rochford, Shoeburyness and Burnham-on-Crouch areas (Figure 13; Lake, Ellison, Hollyer and Simmons, 1977). The depressions are believed to have been cut during the Anglian glaciation and the infilling deposits are probably essentially of late-Anglian to Hoxnian age. Subsequent erosion and aggradation have complicated the sequences preserved within the channels, and brickearths and terrace deposits commonly now overlie varying horizons of the original fill (see Figures 14 and 15). A similar set of 'offshore' buried channels, some at least of which are thought to have been eroded and filled in post-Hoxnian times, has been identified below Flandrian marine and estuarine alluvium. For convenience these are described in Chapter 7, together with the Flandrian deposits that overlie them.

THE ROCHFORD CHANNEL   Boreholes to the west and north of Rochford have proved a relatively steep-sided depression eroded into London Clay. Contours on the London Clay surface in this general area are shown in Figure 13, while Figure 14 gives cross-sections drawn along and across the axis of the channel. The channel has an even basal profile along its axis and a 'U'-shaped cross-profile. Its western edge is particularly steep, being about 1 in 4. London Clay slopes as steep as this are potentially unstable under temperate climatic conditions, suggesting that the channel may have been cut and partly filled in a dominantly periglacial environment. The base of the channel lies at about OD.

The depression is filled with soft, dominantly argillaceous deposits overlain for the most part by loams which are believed to form part of the First and Second terraces of the Crouch. An apparent southward continuation of the channel towards central Southend—here termed the Southend Channel—is filled with up to 14 m of sandy gravel, the upper parts of which form part of the Third Terrace of the Crouch (see Figure 13, p.34). Rapid lateral variations of lithology within the upper part of the sequence within the channel may be related to reworking of the original fill during the formation of the terraces or to subsequent solifluction in areas such as those around Grays, north of Rochford, and near Cherry Orchard Lane [858 896], where the present topography rises above the general level of the Crouch Second Terrace (i.e. 15 m above OD).

Although considerable variation has been observed within the channel-fill, these beds, where thickest, can be divided into the following two lithological units:

B   Clay, silty, greenish grey to brownish grey, soft to firm, with intercalations of silty sand; carbonaceous inclusions and local shell detritus; calcareous 'race' concretions.

A   Sand, silty with subordinate gravel.

The relative thicknesses of these two units varies considerably. Neither unit crops at surface; the areal extent of Unit B beneath the First and Second River Terraces is shown on the 1:50 000 map. The clays of Unit B contain sandy intercalations which demonstrate small-scale channel structures, cross-bedding and fine planar lamination. As a whole the unit tends to fine upwards, and bedding structures have been observed within it at levels of up to 22 m above OD. Some of the clay beds closely resemble the London Clay in their lithology and their consolidation characteristics, and may be derived from its erosion. There is evidence to indicate that, in many places, Unit B passes laterally into silty sands, sands and gravels at the margins of the channel, but

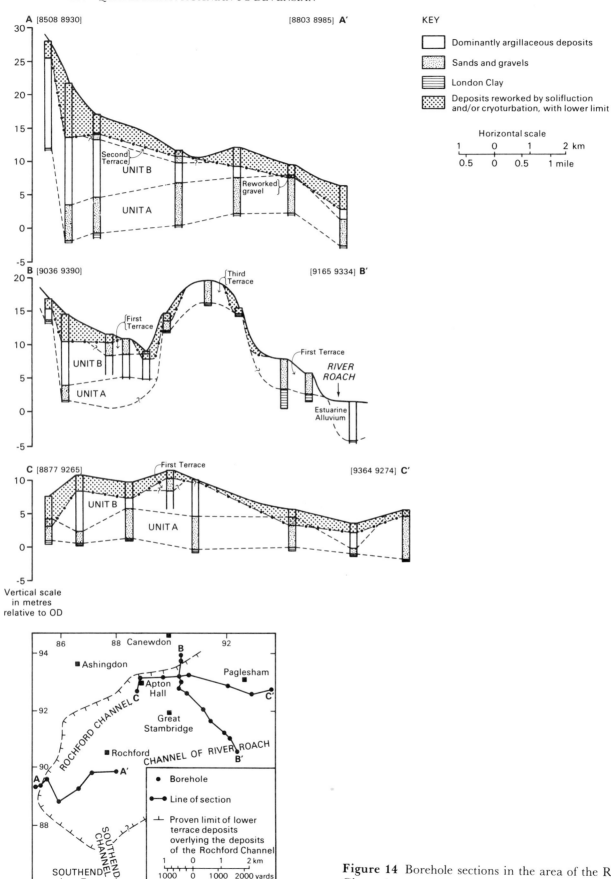

Figure 14 Borehole sections in the area of the Rochford Channel

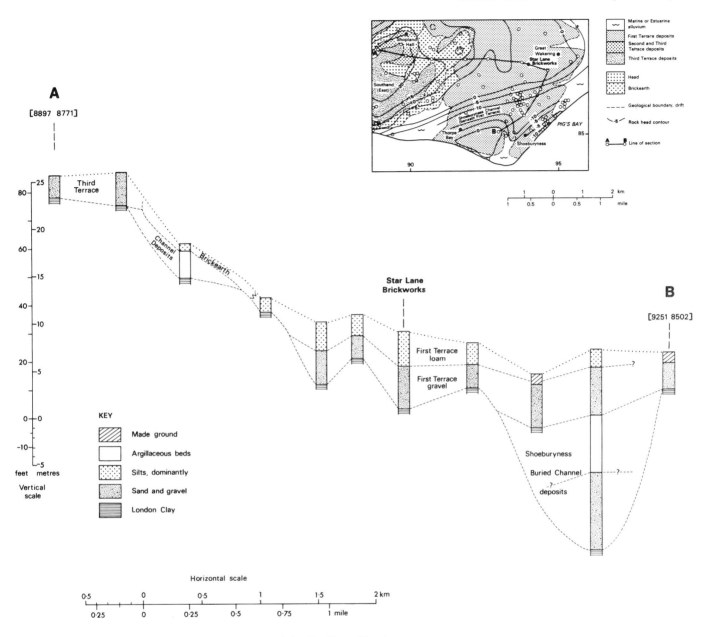

Figure 15 Section through the drift deposits of the Barling–Shoeburyness area

commonly, as around Rochford and Paglesham, it is not practicable to separate the channel fill from later reworked terrace sands and gravels.

Mr Hughes reports that Unit B contains *Cyprideis torosa* (Jones) in abundance. The fossiliferous beds lie at broadly consistent levels ranging from 5.7 to 8.7 m above OD. Brackish water foraminifera are recorded towards the top of these beds and include *Elphidium selseyense* (Heron-Allen and Earland). Pre-Flandrian molluscs were also recovered from a trial pit in this sequence [9039 9296]: they included *Corbicula fluminalis* (Müller) and *Hydrobia ventrosa* (Montagu) group (Kerney 1971, p.92 and personal communication). The fauna suggests a progressive change during deposition from freshwater to cool brackish water conditions, and shows that the channel probably had only a restricted access to the sea.

At greater heights, ranging from 8.4 to 15.7 m above OD rootlet horizons and purplish grey peaty 'soil beds' (see Figure 16), mark the colonisation by plants of the silted-up channel. The presence of these dark mottled organic clays enables the sediments of the channel-fill to be distinguished from more recent, but otherwise lithologically similar deposits.

SHOEBURYNESS CHANNEL   Water wells and site investigation boreholes north of Shoeburyness have proved the presence of another, similar channel system (see Figure 15), beneath the Crouch First Terrace, although the documentation of the sequence is less complete than at Rochford because many of the drillers' logs have confused the argillaceous beds of the channel-fill with London Clay. The shape of the Shoeburyness buried channel and the nature of the deposits within it appear to be very similar to those of the Rochford Channel, though the base of the Shoeburyness Channel reaches a greater depth (15 m below OD).

A well at Thorpe Bay Laundry Company [9188 8567], with a surface level 6 m above OD, proved the following succession:

	Thickness m	Depth m
Made ground	0.51	0.51
Sand and clay	1.53	2.04
Sand and gravel with clay beds and a shell bed	16.76	18.80
LONDON CLAY	—	—

Crouch First Terrace deposits and the buried channel deposits are not separable in this record, but the terrace deposits are likely to comprise only the upper 4 to 5 m. Other wells at Elm Road [e.g. 9380 8553] proved 21 m of alluvial deposits (see Figure 15) overlying London Clay. To the east, wells near Pig's Bay proved 5 to 6 m of sandy clay (Crouch First Terrace loam resting on channel deposits), on sands and gravels 11 to 13 m thick (basal channel deposits), on London Clay. The base of the channel was proved at depths between 10 and 15 m below OD.

Many site investigation boreholes have penetrated the channel deposits, consisting of soft to firm grey silty clays, lying beneath the Crouch First Terrace deposits; they have, however, not established the presence of the lower gravels. The channel clays are distinguished from London Clay by their more obvious silt laminae, by their containing local shell and peat detritus, and by their greater variation in colour, ranging from pale to dark grey.

Taken overall, the limited borehole evidence suggests that there is a bipartite depositional sequence within the Shoeburyness Channel similar to that in the Rochford Channel, and that the sequence at Shoeburyness also shows considerable variations in the relative thicknesses of the two units.

BURNHAM-ON-CROUCH CHANNEL   In the Burnham-on-Crouch area one borehole proved the existence of buried channel deposits beneath the Crouch First Terrace (Simmons, 1978). This borehole [9583 9692], sited near Damner Wick, proved:

Surface level 6 m above OD	Thickness m	Depth m
TERRACE DEPOSITS	2.0	2.0
CHANNEL DEPOSITS		
Clays, silty, olive grey, with race, locally carbonaceous	3.7	5.7
Sand and gravel	5.1	10.8
LONDON CLAY	—	—

The base of the channel deposits lies at 4.7 m below OD, whereas in a nearby borehole [9629 9617], where bedrock was proved at 1.9 m below OD, the drift sequence consisted dominantly of gravelly sands. These sandy beds are apparently composite in origin and include both Crouch Terrace 1 and lower buried channel deposits.

In 1975, the Southminster gravel pits were extended to expose buried channel deposits overlain by cryoturbated Crouch Third Terrace gravels (see p.32). Argillaceous deposits similar to those of the upper (argillaceous) channel-fill have also been recorded adjacent to Crouch Third Terrace gravels near Tillingham, north of the area being described. This suggests that the Burnham buried channel continues in a northerly direction (see Figure 13).

A pit [9074 8741] at Rebels Lane, Sutton proved 3 m of carbonaceous silty clays beneath Head deposits (see Figure 15). These clays may represent further channel deposits preserved at a higher level (up to 17.7 m above OD).   RDL

Age relationships

From the observed relationships with the Crouch Third

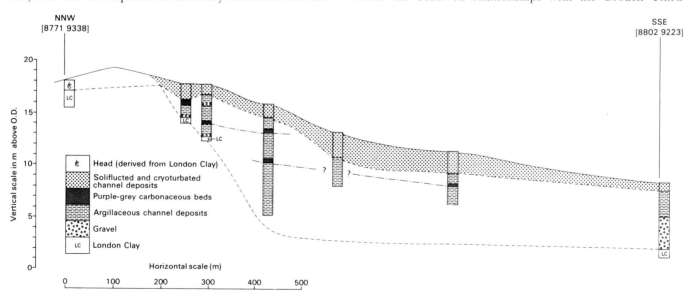

Figure 16 Section along Hyde Wood Lane near Ashingdon in buried channel deposits

Terrace deposits, which are equated altimetrically with those of the 'Boyn Hill' Terrace (see also p.32), it is inferred that these buried channel deposits are mainly of Hoxnian age. They are, therefore, related in time to the channel sequence at Clacton (Pike and Godwin, 1953; Warren, 1955), which is apparently composite in nature and fills multiphase channels that extend down to at least 3.7 m below OD (see Warren, 1955, figure 2). The flora of the upper part of the freshwater sequence reflects a warm temperature climate whereas the succeeding estuarine beds indicate a period of declining warmth (Pike and Godwin, 1953). The faunal studies carried out on the Rochford buried channel suggest that these deposits were laid down in a dominantly cooler climate and probably an earlier phase of aggradation.

Whether glaciation was a factor in the development of the buried channels is uncertain, but the hollows have probably been eroded under glacial or near glacial conditions (Lake and others, 1977). The drainage pattern, particularly the breaching of the London Clay ridge between Hadleigh and Southend by the Southend Channel (see p.34), the bow-like course of the Rochford Channel and its apparently 'blind' 'blind' eastern side south of Rochford, the northwards course of the channel drainage in the Burnham area and beyond, and associated Crouch Fourth Terrace spreads all indicate that the rivers could not flow freely to the east.

Further Pleistocene deposits which are possibly comparable with those detailed above are present within buried channels beneath Flandrian alluvium and are described separately (see p.47–52).   RDL, GWG

Details

ROCHFORD CHANNEL   Sections in the Prittle Brook, west of Temple Farm [884 884] typically showed the following succession:

	Thickness m
ALLUVIUM	
Loam, brown, silty	1.0 to 1.8
Gravel	0.5 to 0.75
CHANNEL DEPOSITS	
Clay, grey and brown mottled (partly augered)	seen 1.7
	GWG

On Ironwell Lane, Rochford [8624 9097], a trial pit sited near the centre of the channel exposed the following sequence:

	Thickness m	Depth m
CROUCH SECOND TERRACE DEPOSITS		
Gravel, fine- and medium-grained, with subordinate coarse gravel	0.7	0.7
CHANNEL DEPOSITS		
Clay, silty, very soft, grey-green, with brown oxidised patches; sharp well-defined base; a pipe infilled with gravel extended from the bed above down to 1.75 m	0.6	1.3
Silt, clayey, firm, pale purplish grey, with some dark purple patches containing carbonaceous traces, rare fine-grained subangular white (bleached) flints; gradational base	0.15	1.45
A fining upwards sequence of laminated medium- and coarse-grained sand with clayey		

	Thickness m	Depth m
lenses, grading up into firm mottled greenish grey and brown clayey silt with patches of fine-grained sand; sharp uneven base with small scale piping	0.45	1.90
Sand, fine-grained, silty and clayey, finely laminated, mottled brown and grey, with some fine- and medium-grained sand lenses, and coarse-grained sand and subangular gravel beds up to 4 cm thick; race clusters occur at 2.95 m	1.05	2.95

The purple carbonaceous clayey silt near the top of the Channel Deposits probably represents an emergent phase during the channel infilling, when plants colonised the mud flats.

North of the Brays–Ballards Gore road, the surface of the river loams rises gradually northwards, and much of the ground is covered by an unmapped veneer of stony brown loam, which overlies the upper argillaceous unit of the Channel Deposits.

Auger holes and pits along the north–south line of Hyde Wood Lane [c. 879 930] proved the sequence illustrated in Figure 16. One pit [8783 9313] exposed:

	Thickness m	Depth m
HEAD		
Clay, silty, stiff, mottled brown, with scattered gravel	1.15	1.15
TERRACE DEPOSITS AND CHANNEL DEPOSITS		
Silt, clayey, firm, mottled brown, with remnant yellow-brown silt and clay laminae from 1.38 m; gradational base	0.45	1.60
Gravel, fine-, medium- and coarse-grained, in a silty medium- and coarse-grained sand matrix; scattered frost-shattered flint shards occur throughout and the gravel constitutes up to 20 per cent of the bulk	0.05	1.65
Sand, medium- and coarse-grained, laminated, iron-cemented	0.07	1.72
Silt, clayey, stiff, brown, with scattered selenite crystals	0.23	1.85
Silt, clayey and silty fine- and medium-grained sand with remnant lamination; gradational base	0.71	2.56
Sand, silty, fine-, medium- and coarse-grained, with some randomly orientated subangular gravel; irregular erosive base	0.08 to 0.16	2.72
Silt, sandy fine- to coarse-grained, and fine- to coarse-grained sand, cross-bedded and ripple-laminated; sharp erosive base	0.48	3.12
Silt and very fine-grained sand, firm, pale grey and brown, finely laminated; sharp undulating base with 0.1 m relief	0.06	3.18
Silt, clayey, firm, pale purplish grey, with darker purple silt in the uppermost 2 to 3 cm; gradually grading down into grey and green-grey interbedded, finely laminated silty clay and clayey silt, becoming gravelly in the basal 20 cm (augered from 3.90 m).	1.97	5.15
Silt, very clayey, firm, dark yellow-brown, with fine- to coarse-grained sand and gravel, particularly at the base	0.55	5.70

	Thickness m	Depth m
LONDON CLAY		
Clay, silty, stiff, brown	0.20	5.90

The beds above the erosion surface at 2.72 m are Second Terrace deposits. At least two phases of deposition and reworking can be recognised, with erosional bases at 1.65 m and 2.64 m.

Along the eastern margin of the upper unit of the Channel Deposits between Southend, Rochford and Southend Airport, there is usually a lateral passage eastwards into silty sands, sands, and, generally subordinate, gravels. Graves in the main part of the Southend Cemetery [around 883 880], open at the time of the survey, showed brown and buff loam, 'brickearth' of the Crouch First Terrace, 1.0 to 1.8 m thick, passing imperceptibly downwards into Channel Deposits of brown fine-grained sand (2 m seen, with a further 1 m augered). A temporary exposure [8788 8803] in the (new) western extension to the cemetery showed that there is a lateral passage into the upper clayey unit in this direction.

At Little Stambridge Hall Lane [8870 9133] a trial pit proved a complex sequence of Head deposits, with cryoturbation structures extending into the underlying laminated silty and fine-grained sandy Channel Deposits.

Pits at Ballards Gore [90 93], an area of disused brickpits and gravel workings, proved Second Terrace gravels overlying buried Channel Deposits. The latter vary in lithology from orange-yellow laminated micaceous silt and fine-grained sand to shelly grey silty clay. *Cyprideis torosa* and *Elphidium selseyense* were recorded from the Channel Deposits in two trial pits [9039 9298; 9028 9277] and the molluscs *Corbicula fluminalis* and *Hydrobia ventrosa* were collected at this first site (see p.29). These faunas indicate a fresh to brackish water environment of deposition. The section recorded in this trial pit [9030 9298] is:

	Thickness m	Depth m
CROUCH SECOND TERRACE		
Sand, medium- and coarse-grained, bedded, with subordinate amounts of gravel, sharp base	2.2	2.2
CHANNEL DEPOSITS		
Sand, very fine-grained, finely laminated, containing coarse-grained sand and lenses of fine-grained sand; sharp base	1.05	3.25
Coarsening upwards sequence from stiff, shelly, black, very silty clay to laminated, greenish grey and brown, shelly, micaceous, clayey silt and micaceous silty fine-grained sand, with small scale cross-stratification in the upper 5 cm. (Augered from 4.95 m)	2.85	6.10
Sand, gravelly (augered)	0.10	6.20

The lowest unit recorded probably represents the upper part of the basal gravelly unit of the Rochford Channel Deposits (see p.27).

RAE, GWG

BURNHAM-ON-CROUCH CHANNEL. The composite section of the Southminster gravel pit [953 982] is as follows:

	Thickness m	Depth m
Clay, sandy, silty, pale greenish grey, brown and orange mottled, with distinct greenish grey vertical mottles; scattered subangular to subrounded flints concentrated in small pockets;	1.5	1.5
Clay, sandy, silty, dark grey to purple, slightly organic; local traces of fine lamination	0.3	1.8
Clay, sandy, silty, pale greenish grey, brown and orange mottled, as item above but with increased green mottling around rootlet traces	0.5 to 1.0	2.8
Sand, fine-and medium-grained, orange and greyish orange; subordinate fine- and medium-grade gravel; cross-bedded	1.3	4.1

Locally pods of gravel which are related to the inferred Crouch Third Terrace bench, overlie this sequence. The flints in the topmost item in the section probably come from these gravel pods and have probably been incorporated and preserved by cryoturbation. Where the brown colouration is dominant the top clay units resembles London Clay. Elsewhere it is massive and unstratified with the columnar jointing that is characteristic of loess.

RDL, RAE

RIVER TERRACE DEPOSITS

The terrace gravels and loams of fluviatile origin are, for the most part, related to the present drainage pattern. Their distribution is confined to ground bordering the main alluvial tracts of the rivers Thames, Roach and Crouch, though between Rochford and Shoeburyness they are more extensive and effectively blanket the Roach–Thames interfluve. Locally they are overlain by Head and Brickearths (see below).

Extensive spreads of Terrace Deposits are present in the south–west of the district between Stanford le Hope and Fobbing; they extend westward into the Romford (257) district, where they can be linked by mapping to the established Lower Thames sequence and are accordingly designated.:

Fourth Terrace = 'Boyn Hill';
Third Terrace = 'Taplow';
Second Terrace = 'Upper Floodplain';
First Terrace = 'Lower Floodplain' (not exposed in the present district).

The absence of terrace deposits between Fobbing and Southend hinders correlation between the Lower Thames terraces and those preserved north and east of Southend. The terraces in this latter area are, therefore, numbered on a separate basis consistent with that employed in the Crouch valley, namely, in reverse order of age, the Crouch First, Second, Third and Fourth terraces. In the following account the two suites of terraces have been described separately in order to avoid confusion between the two numbering systems. The Crouch Third Terrace in the Southend district is provisionally thought to be equivalent to the Fourth Terrace of the Lower Thames (Lake and others, 1977).

The terrace deposits of south-east Essex have recently been studied by Bridgland (1980; 1983) who recognised a suite of high level flinty gravels embracing the Crouch Fourth Terrace and the Sand and Gravel of unknown origin (see p.25): these deposits are characterised by the predominance of materials derived locally and from the Weald. His low-level suite comprises the Crouch First to Third Terraces and contains some significant exotic material, including vein quartz, ortho- and metaquartzite, Carboniferous chert, *Rhaxella* chert and igneous rocks. He took the presence of *Rhaxella* chert to indicate an Anglian or post-Anglian age, considering that this material was introduced by Anglian ice. Bridgland (1983) correlated the Crouch Third Terrace (his Southchurch Gravel and Asheldham Gravel) with the Boyn Hill Terrace (as above).

Throughout, considerable difficulty has been experienced in trying to allocate unstratified or poorly stratified silts to an appropriate genetic category. Some of these silty deposits are fluvial overbank or channel plug deposits and can be described as loam (River Brickearth); others are columnar-jointed and show the grading characteristics of loess, so that they are best described as Brickearth. However, a large proportion of the silty deposits lack any diagnostic characters and cannot be allocated with certainty to either category. Where they are associated with river terraces, they have been mapped as loam (River Brickearth), but where they are above terrace level, they have been mapped as Brickearth. This arbitrary procedure may have led to some loessic deposits, with their associated stability problems, being classed as terrace loams.

FOURTH TERRACE DEPOSITS   These form only a single spread, which lies in the south-west of the district, to the north-west of Fobbing. Further isolated patches may be present beneath the Head Deposits around Corringham. The present surface level of the spread ranges from 18 to 32 m above OD. Field relationships suggest that the deposits have an irregular base and hence a variable thickness, possibly because they were deposited during more than one cycle of aggradation. This may also account for the rather wide range in altitude of the spread. The limited information relating to the composition of the deposit suggests that it consists of discrete beds of sandy gravel within a silty sand sequence. A borehole [7171 8409] located between the outcrop of this terrace and the Third Terrace, east of Fobbing, proved 14.9 m of silty sands with subordinate bands of silty clay and gravel beneath 1.0 m of Head: this thickness is regarded as exceptional.

THIRD TERRACE DEPOSITS   There are extensive outcrops of the Third Terrace between Fobbing and Stanford le Hope. A back-feature is locally present at about 18 m above OD and the surface of the deposits ranges from 9 m above OD up to this height. The outcrops are almost entirely surrounded by younger Head deposits at the surface.

In composition the terrace deposits consist for the most part of sandy flinty gravels, which range in thickness from 2.4 to 5.0 m over much of the outcrop. Limited borehole evidence suggests that the base of the gravels rests on a bench lying at c.9 to 13 m above OD.

SECOND TERRACE DEPOSITS   Patches of Second Terrace gravels are preserved around Mucking. They extend southwards and westwards through Gobions to West Tilbury in the Dartford (271) district. The topographically higher parts of these deposits are generally covered by Head, but a back-feature for this terrace is recognisable between 7.5 and 10 m above OD.

The deposits range in composition from sands, with very subordinate sandy gravels and fine clay partings, to sandy gravels. Gravel workings in the Mucking area expose up to 3.5 m of these deposits beneath 1.8 m of Head. Records of wells at Gobions and West Tilbury (Curtis and others, 1965) indicate that the base of the deposits lies at between 1.5 and 4.6 m below OD.   RDL

CROUCH FOURTH TERRACE DEPOSITS   Small patches of Crouch Fourth Terrace are present north of the Crouch valley in the Althorne area and south of it at Canewdon. Surface levels range mainly from 27 m to 42 m above OD; more than one terrace may be represented.

Further, more extensive, occurrences are present near Eastwood, near Hadleigh golf course, and in the west of Southend where they cap elongate ridges. The present surfaces slope down gently eastwards, with levels ranging widely from 51 m to 33 m above OD. Borehole records in this area indicate that the deposits range in composition from sandy gravel to gravelly clay. Because of this they are locally very difficult to separate from Head. Their thicknesses also vary from 0.6 to 3.5 m, and temporary exposures, which show that the gravel lies in pockets let down into the London Clay, confirm that these variations result from intense cryoturbation.

CROUCH THIRD TERRACE DEPOSITS (loams and gravels)   Small patches of Crouch Third Terrace gravels are present in the Crouch valley at South Green and extend eastwards from near Shotgate to near Burnham-on-Crouch. Except in the extreme west, where surface levels are around 40 m above OD, the maximum surface levels of these deposits generally range between 15 m and 24 m above OD.

Locally in the Burnham-on-Crouch area, surface cappings of silty loam up to 2.4 m thick overlie the gravels. These are present at levels generally above 15 m above OD ; localised lower occurrences have probably been solifluced downslope. The underlying terrace gravels range up to 4.5 m in thickness. They have a size range from sandy gravel to clayey sand. Locally, significant clayey intercalations are present. A good terrace flat is present near Stoneyhills [955 985], but the only deposits preserved on it are of gravelly materials which have been let down into the underlying buried Channel Deposits by cryoturbation.

An isolated occurrence of Crouch Third Terrace loam, at a height varying from 13 to 19 m above OD., is present near Great Stambridge, north-east of Rochford, where it is surrounded for the most part by Head deposits. The loam is apparently somewhat disturbed as it has no clearly defined basal contact. Generally, however, there is a tendency in boreholes for these deposits to coarsen downwards from a sandy silty clay to a silty sand with a variable gravel content. The basal bench of the terrace here ranges from 11 to 16 m above OD.

A pit section near Stambridge Sewage Works [9118 9203] showed, beneath disturbed ground, a threefold sequence of sand and gravel, on loam, on sand and gravel; the sequence may indicate either the presence of two cycles of deposition, or deposition by a meandering river.

Loams and gravels ascribed to the Crouch Third Terrace occur in the eastern and central parts of Southend. The surface level of these deposits ranges from 16 m to 34 m above OD. Some of these deposits may possibly be composite in origin, having been derived by solifluction from the Crouch Fourth Terrace at west Southend, and reworked near Shopland Hall [899 882] whilst the Crouch Second Terrace was being deposited; these possibly later deposits cannot be mapped out separately from those of the undisputed Crouch Third Terrace however. A more characteristic range of levels of the latter is from 20 to 28 m above OD. In certain areas, where borehole information is sufficiently dense, a distinct sub-terrace bench is recognisable with a level varying from 17 to 23 m above OD (Figure 13).

Details of the terrace deposits are known mainly from a wealth of borehole data in an area extending northwards from the central part of Southend to Prittlewell. In most of this area the upper surface of the terrace is mantled by uniform buff loam ('brickearth') to a depth of about 2 to 4 m, though a maximum of about 5.5 m has been recorded locally near Southend Central Station. Where a sub-terrace bench can be recognised, some 6 to 8 m of sand and gravel intervene between the 'brickearth' and the underlying London Clay.

The most striking feature of the terrace is the associated gravel-filled, slightly sinuous, depression with small tributary feeders which extends northwards from the area of the Central Station to Prittlewell. This is here named the Southend Channel (Figure 13). It is about 0.8 km wide and slopes northwards at a gradient of about 1 in 16, with channel sides sloping at about 1 in 10. Before the postulated removal by erosion of the 'brickearth' and the immediately underlying gravel in the central, deepest part of the channel, the maximum thickness of sandy gravel may have been about 19 to 20 m. At Prittlewell the channel floor appears to be graded to the base of the Rochford Channel (see p.27). The Southend Channel appears to have been cut by a large river that drained northwards across the Hadleigh-Southend London Clay ridge, possibly whilst the drainage farther east was blocked (see p.31). The main course of the Rochford Channel was probably also established at this time, and hence much of the sand and gravel at its base may be contemporaneous with the gravel-fill of the Southend Channel. Subsequent aggradation of the two channels presumably continued until the eastward course of the main 'Thames' river was established, sea-level rose, and the Crouch Third Terrace was laid down. Exposures of the sands and gravels of this terrace can still be seen in a number of disused gravel pits in the Southend area.

Further small patches of Crouch Third Terrace gravels are present at Beauchamps [908 885] and near Barrow Hall Farm [920 880], with surface levels approaching 15 m above OD. Borehole evidence suggests that this terrace continues south-westwards from Barrow Hall Farm beneath Head deposits, with a basal bench ranging from 12 to 15 m above OD.

GWG, RDL

CROUCH SECOND TERRACE DEPOSITS (loams and gravels) Small remnants of Crouch Second Terrace gravels are present in the Crouch valley west of Hullbridge. A further occurrence at Burnham-on-Crouch forms a composite feature with the Crouch First Terrace (see below). As a consequence, the present surface level of the deposits ranges from 5 to 27 m above OD; it generally increases upstream.

Within the Southend-Paglesham area, Crouch Second Terrace deposits are restricted to the areas north of Rochford, east of Great Stambridge, and around Lambourne Hall, near Canewdon.

North of Rochford these deposits form a relatively high plateau at 11 to 15 m above OD, and have a basal bench at about 7 m above OD. They have been worked between Rochford and Doggetts where they are about 3 to 4 m thick. The base of the terrace was seen in a sewage trench in the Doggetts area where the contact with the underlying London Clay was extremely irregular and cryoturbated. In an old pit [880 915] south of Doggetts the surface loam penetrates the underlying well-bedded sand and gravel in frost wedges, spaced at intervals of 10 to 15 m along the old working face. Patterned ground with fossil ice-wedge polygons is present in the adjacent fields and can be identified on aerial photographs.

Up to 5 m of sand and gravel are generally present in the two outcrops near Lambourne Hall, where the surface level ranges from 10 to 12 m above OD. The larger and more northerly occurrence has been worked extensively, and there are conspicuous, fossil ice-wedge polygons on the undisturbed ground surface. A basal bench for these deposits has been proved by drilling to lie at about 6 to 7 m above OD.

South of the River Roach, remnants of the Crouch Second Terrace are few, although a composite sequence of gravels ascribed to the Second and Third terraces is present near Shopland Hall [899 882].

In the Eastwood area [855 886], an outcrop of gravel has been tentatively allocated to the Crouch Second Terrace because of its relatively high topographical level of about 15 m to 18 m above OD. In its western exposed part, however, it falls north-eastwards to below 7 m above OD and it is composite (see p.35). From borehole and surface evidence the gravel, which is rarely more than 1 to 2 m in thickness, appears to extend northwards and eastwards some considerable distance beneath the alluvial loam ('brickearth') designated as Crouch First to Third Terrace, though it does not extend as far as the Airport. Over most of this area it overlies the deposits of the Rochford Channel (see p.27; Figure 14). The gravel is exposed near its northern feather-edge on either side of the Roach near Hawkwell Hall, where the stream has cut down into the underlying Rochford Channel deposits, but it is too thin to map.

GWG, RDL

CROUCH FIRST TERRACE DEPOSITS (loams and gravels); (including Crouch First to Third Terrace loams in the area of the Rochford Channel) With the exception of three relatively small patches on either side of the River Crouch at Burnham-on-Crouch and further occurrences east of Stoneyhills, the main spreads of Crouch First Terrace Deposits lie between Paglesham and Southend. The largest outcrop at Burnham-on-Crouch is composite and includes gravels of the Crouch Second Terrace. It is overlain by 1 to 2 m of brickearth in one area. The surface level of this deposit generally ranges from 2 m—the level of present day alluvium—to about 6 m above OD, but is at 9 m above OD in the area where the First and Second terraces merge. Limited borehole evidence suggests that 2 to 5 m of gravelly sand are present, both in the outcrop area and also beneath Head to the north, with a basal bench ranging from 2 m below to 4 m above OD. In the south-east around Burnham Wick, the terrace deposits overlie the Burnham buried channel deposits which are clayey in their upper part (see p.30).

North of the River Roach in the Paglesham area, there are outcrops of Crouch First Terrace deposits which have affinities with both the deposits to the south and the channel deposits to the west (see p.27). Boreholes sited on the dissected sheet of upper loam proved up to 1.4 m of massive, unstratified sandy silt and silty clay overlying gravelly sands which vary in thickness from 0.2 m to 6.2 m. The variation in thickness of the gravels is related in part to the varying depth of bedrock, which ranges from 3.8 m below to 3.1 m above OD (see Figure 13).

South-east of Paglesham a single borehole [9289 9255] has proved 4.5 m of soft silty clay overlying London Clay and it is suspected that other intercalations of clayey material within the gravel deposits of this area may occur, particularly where the bedrock is comparatively deep (see Figure 13).

In the area of the Rochford Channel (p.27; Figure 14) the terrace deposits have been designated Crouch First to Third Terraces, and have been shown throughout as loams (River Brickearth) on the 1:50 000 sheet. This has been necessary because of the difficulty in separating the younger terrace loams from the silts of the Rochford buried channel and their varied solifluction products. Along the western and southern limits of the area, a thin, cryoturbated gravel spread separates buried channel silts, below, from the younger terrace loams, above. An outcrop of this gravel has been mapped at Eastwood, where it has been tentatively assigned to the Crouch Second Terrace because of its height (see p.34). East of Prittle Brook a thin loamy gravel separates the First Terrace loam from the Channel Deposits, but its relationship to the gravel further west is not clear; it has been classified as Crouch First Terrace because of its somewhat lower level. Typical unstratified columnar-jointed loams, which are usually subdivisible into calcareous and non-calcareous parts (see below), mantle most of the area on the south side of the Roach and also occur from Brays westwards as far as the railway. East of Brays and in the valley of the Roach between Rochford and Hawkwell Hall, the terrace deposits comprise rather clayey loams with a variable but thin cover of solifluction products (not mapped separately).

North of the Roach south-west of Great Stambridge and between Barling and Shoeburyness, there is an extensive spread of loams and gravels which are mainly part of the Crouch First Terrace. Locally surficial mass movement has made it difficult to separate the various deposits. Yellow-brown sandy silts are dominant at the surface and have been extensively worked for brick-making. They are generally very difficult to distinguish from Brickearth and Head Brickearth (p.38), and form an irregular sheet up to 4 m thick on the gravels of the First Terrace. These loams, as elsewhere, (p.38) may commonly be divided into upper, non-calcareous, and lower, calcareous, parts. Both show evidence of extensive rootlet penetration; calcareous root-cores and calcareous 'race' nodules, which commonly reach 1 cm in diameter, are common in the lower part. The base of the non-calcareous part is generally, though not invariably, sharp; very locally in the Cherry Orchard Lane area (p.37) a dark horizon approximately separates the two and has been termed a 'parabraunerde' (i.e. an inter-stadial soil) by Gruhn, Bryan and Moss (1974, p.65). more usually, however, the non-calcareous/calcareous interface is everywhere at about the same depth below surface. This, combined with the lithological similarity of the two subdivisions, suggests that the absence of lime in the upper division is the result of leaching by ground-water.

The loam deposits are generally unstratified. There are rare stringers of pebbles and isolated pebbles in places, and traces of primary stratification, possibly indicating sub-aqueous deposition, are locally evident in the lower part: because of this the silts were mapped with the underlying terrace deposits. It is however, quite possible, that the deposits mapped as loams may include loessic material, especially as in many places around their western and southern limits (e.g. in the Stroud Green area, between Cherry Orchard Lane and Eastwood, between Sutton and Bournes Green), the loams pass, both laterally and upwards, apparently without break, into undoubted Brickearth and Head Brickearth deposits which mantle the adjacent slopes.

The contact between the silts and the underlying gravels is apparently gently undulating on the broad scale. Locally the gravels are absent so that the silts rest directly on London Clay.

In the Barling and Shoeburyness area the Crouch First Terrace gravels range from silty sands to sandy gravels. Locally, they crop out beside the alluvial flats. The gravels generally range in thickness from 1 to 2.5 m, although exceptionally in a borehole [9175 8953] near Mucking Hall a thickness in excess of 6 m has been recorded.

In this same area the surface level of the deposits assigned to the Crouch First Terrace, again composite in nature, ranges from about 2 m to 15 m above OD; Generally the base of these terrace deposits (see Figure 13) falls gently away to the north, east and south from a height of about 6 m above OD, centred on Blue House Farm [917 887]. The highest recorded level of 7.1 m above OD was in a borehole [9020 8909] near Butler's Farm, whereas levels below OD were observed to the east of Great Wakering and near Barling Hall [936 898]. RDL, GWG

Fauna of the Terrace Deposits

An antler fragment of *Megaceros sp* is recorded from gravels near Prittlewell by Gruhn, Bryan and Moss (1974, p.60). The molar of a mastodon from nearby, possibly reworked, gravels on Hobleythick Lane was also noted. The same authors (1974, p.63) record the following fauna from Baldwins Farm, Barling (in Crouch First Terrace Deposits): *Elephas primigenius, Equus, Cervus elaphus, Bos primigenius, Bison, Megaceros* and '*Rhinoceros*'. Gravels near Thorpe Hall yielded a tooth of *Elephas antiquus*. The authors compare the fauna with that described from Ilford by Sutcliffe (1964) which he assigned to the Ipswichian. At Barling, Middle Acheulian implements are associated with the fragmentary bone material. The presence of this relatively warm fauna in the terrace supports a correlation with the Ipswichian, though it is possible that some of the faunal remains may have been derived from earlier deposits.   RDL

Details

FOURTH TERRACE DEPOSITS   In a former gravel pit [713 845] near Fobbing Waterworks, H.G. Dines (in MS. 1923) recorded:

	Thickness m
Brown slightly loamy clay	0.9
Clean, cross-bedded sand	2.1
Gravels, bedded ('dense')	3.0

This pit has been worked out and is now degraded, but field relationships nearby indicate that the base of the gravels is highly irregular in nature.

THIRD TERRACE DEPOSITS   In 1923, H.G. Dines observed the following section in the now backfilled Abbott's Hall gravel pit [6928 8275]:

	Thickness m
Brown loamy clay	0.3 to 0.9
Sand and gravel, light coloured, cross-bedded, White sand beds up to 0.3 m or more thick; subordinate thin beds of fine gravel; seams or lenticles of brown clay about 0.03 m thick	2.4 to 3.0
Sand, soft, unbedded with a basal layer of large relatively unworn flints	0.6 to 0.9

LOWER LONDON TERTIARIES (see p.10)   —

SECOND TERRACE DEPOSITS   Second Terrace Deposits have been extensively worked in the Mucking area. A section in the Wharf Road pit in 1972 [6936 8155] showed beneath 2.4 m of Head deposits:

	Thickness m
Sands, fine- to medium-grained, ochreous, well laminated. Cross-bedding and ripple structures; persistent clay partings to 0.3 m thick	1.7
Gravel, sandy, fairly well sorted	0.6

This sequence passed westwards by interdigitation into sandy gravels. Other sections in the western part of the pit showed up to 3 m of poorly bedded flint gravels with subordinate sand lenses and irregular bands of ferruginous staining.   RDL

CROUCH FOURTH TERRACE DEPOSITS   Two patches of terrace deposits, which are elongated in an east–west direction and slope gently eastwards, have been mapped near Leigh-on-Sea. In the northern spread, augering showed the deposits to be variable, ranging from stony loam, through coarse- and fine-grained sand, to gravel. Trial holes on the site of Leigh Fire Station [849 873] record soil and disturbed ground 0.6 m, on gravel ranging from 1.4 to 2.1 m in thickness, overlying London Clay. A patch of Head, about 0.8 km east of the eastern end of the main terrace was probably derived from this terrace.

The southern spread of terrace occupies largely built-up ground, but a few temporary sections and boreholes show the deposits to be mainly rather clayey gravel, rarely more than 1 to 2 m thick. At the eastern extremity of the terrace south of Cliff Avenue and west of Balmoral Road [873 860], numerous site-investigation boreholes record very variable deposits ranging from 'silty clay' to 'dense

gravel', overlying London Clay and extending from 1.0 m to 2.5 m below ground level (including made ground). The apparent variability may be due to cryoturbation, for exposures in an extension to the nearby Telephone Exchange [8742 8586] showed marked deformation of the gravel/London Clay contact.   GWG

CROUCH THIRD TERRACE DEPOSITS   At Stambridge Sewage Works, a pit [9118 9203] proved:

	Thickness m	Depth m
MADE GROUND	0.6	0.6
Sand, gravelly, bedded; the sand comprised 60 per cent of the bulk, being mainly coarse-grained; the gravel comprised subangular (60 per cent) and subrounded (40 per cent) flints; sharp base with 10 cm relief	0.65	1.25
Sand, fine-grained and clayey silt, micaceous, finely laminated, soft to firm, mottled dark orange-yellow, with some vertically oriented reduced veins; sharp base	0.80	2.05
Sand, coarse-grained, gravelly	0.45	2.50
		RAE

Around Southend the sand and gravels of the terrace, tend to be preserved near the Southend Channel. They were extensively worked in the past in Southend and Prittlewell and a number of much degraded, working faces can still be seen in minor exposures.   GWG

In the Sutton area a trial pit [8915 8747] near Jordans showed the following section:

	Thickness m	Depth m
Terrace loam		
Silt, clayey, firm, brown, with scattered fine-grained angular gravel; gravelly stringer at sharp irregular base	1.35	1.35
Silt, clayey, firm, greyish orange, with abundant calcareous root-cores and scattered 'race' nodules; fine- to medium-grained flints scattered throughout; irregular cryoturbated base	0.30	1.65
Silt, sandy, clayey, firm, mottled yellow-orange, with small (2 cm maximum diameter) irregular patches of brown silty clay; 'race' nodules up to 3 cm diameter	0.70	2.35
Clay, silty, firm, dark grey, with irregular patches of brown fine-grained sand; discontinuous bed; cryoturbated base with small veins of soft fine-grained sandy clayey silt extending into the underlying bed; scattered 'race' nodules	0.15	2.50
Silt, fine-grained, sandy, soft, pale grey, with scattered 'race'; becoming mottled downwards and grading into the bed below	0.25	2.75
Silt, very clayey and silty clay, firm, brown and pale grey, with 'race'; scattered pebbles from 2.8 to 3.1 m and from 3.6 m to base	1.15	3.90
Terrace gravel		
Sand and gravel (augered).	0.70	4.60

The beds between 1.65 and 2.50 m show evidence of solifluction and thus illustrate the composite nature of the sequence.   RAE

CROUCH SECOND TERRACE DEPOSITS   There are several poor sections around the degraded margins of the disused and partially filled Creeksea Sand and Ballast Pit [920 940] near Canewdon. Fossil ice-wedge polygons filled with loam were extensive in the terrace gravel deposits, but most were destroyed during excavation and few are still visible. Up to 1.2 m of alluvial loam, comprising brown firm sandy clayey silt with scattered flint pebbles, formerly covered much of the area; it is now poorly exposed in the pit sides. A section [9200 9400] in the underlying sand and gravel showed:

	Thickness m
Gravel, fine- to coarse-grained, forming coarse bedded units set in a medium-grained sand matrix, and grading upwards in places into thin beds of medium- to coarse-grained sand	2.0
Sand, medium- to coarse-grained, with some medium- to coarse-grained gravel lenses and interbeds	2.5

Immediately to the south, similar lithologies and thicknesses are present in a pit at Loftmans [9165 9395].

Up to 1.7 m of loamy deposits form a capping to the sandy gravels of the Crouch Second Terrace to the south-east of Doggetts: a thin (0.16 m) calcareous band lay at the base of the loams in one borehole [8799 9177]. The ground surface here shows traces of fossil ice-wedge polygons (p.34). Up to 1.5 m of fine- to coarse-grained sands, interbedded with fine- to coarse-grained gravels with a sandy matrix, were noted below 1.6 m of clayey silts in a degraded gravel pit nearby [8802 9152].   RDL, MRH

CROUCH FIRST TERRACE DEPOSITS (including Crouch First to Third Terrace loams in the area of the Rochford Channel)   The surface loams (brickearth) have been extensively worked for brick-making in the Cherry Orchard Lane area. The best section is given below; it was recorded in 1972 from an old pit [856 896] south-west of the area then being worked.

	Thickness m	Depth m
Silt, pale brown, with rootlet traces; base fairly sharp and even	1.1	1.1
Silt, dark brown, with scattered flint pebbles; base very sharp and even	0.3	1.4
Silt pale brown with abundant calcareous root tubes (floor of pit formed by sandy gravel)	1.3	2.7

A borehole [8552 8959] 50 m to the west continued the section downwards as follows: sandy gravel 1.0 m; Rochford Channel silts, silty sands and silty clays 8.6 m; sandy gravel and gravelly sand 5.3 m; London Clay (upper surface at 0.5 m below OD). The working faces of the pit [c 856 898] show from 1.8 to 2.3 m of columnar-jointed brickearth of which the top 1.4 to 1.7 m is non-calcareous. The junction with the underlying calcareous brickearth is generally fairly sharp, but is only locally marked by the dark layer (parabraunerde) noted above (p.35). Analyses of the silts by Gruhn, Bryan and Moss (1974, p.64) indicate that a loessic origin for them is possible.

A section [8650 9025], which was exposed in 1972 in the drainage sump on the north-east corner of the brickworks, showed about 0.6 m of buff, non-calcareous brickearth passing down, without a break, into a similar thickness of calcareous brickearth, which in turn overlay 0.5 to 0.6 m of rather silty gravel resting on brown sandy silt (Rochford Channel deposits). The gravel is absent in a borehole [8572 9013] 400 m to the WNW of this section, where the loams pass directly down into the silts of the channel deposits (see below).

Between the brickworks and the Eastwood area to the south-west, the western part of the terrace gravel and the underlying deposits were well exposed in boreholes and in a trench. The gravels are

interpreted as Crouch Second Terrace (p.34). They thin out westwards, so that the brickearths directly overlie the underlying Rochford Channel deposits, which are banked against a steep slope of London Clay draped with clayey Head up to 2.5 m in thickness.
GWG

Near Cherry Orchard Lane, a trial pit [8526 8934] near to the western edge of the channel deposits showed:

	Thickness m	Depth m
Silt, clayey, firm to hard, brown, faintly laminated; less clayey and containing calcareous roots and small 'race' nodules below 1.15 m, with subrounded and subangular fine- and medium-grained gravel stringers	2.00	2.00
Silt, clayey, dark brown, with scattered flints (?'parabraunerde')	0.07	2.07
Silt, clayey, orange-brown with calcareous roots and 'race'; becoming softer with depth; some fine-grained gravel and coarse-grained sand from 4.2 to 4.5 m (augered from 4.00)	4.78	6.85

RAE

The loams extend westwards up the hillslope beyond the limits of the Crouch Second Terrace and the Rochford Channel where they have been mapped as part of the Brickearth that mantles the slopes to the west (p.38).
GWG

The Barling Hall Sand and Gravel Pit [9355 8965] exposed Crouch First Terrace loam (with a non-calcareous/calcareous differentiation) overlying thick sand and gravel also assigned to the terrace deposits. Fossil ice-wedge polygons were prominent within the loam; ice-wedge casts formed free-standing walls in the worked-out areas and were exposed in the pit sides. The underlying London Clay was deformed into ridges up to 0.3 m high beneath the ice-wedges. A section in the terrace loam [9345 8967] is as follows:

	Thickness m
Silt, clayey, sandy, yellow-brown with vertical fissuring	1.5
Silt, sandy, orange-brown, with calcareous cemented root tubules, scattered pebbles and strong vertical fissuring; the pebbles commonly have vertical long axes indicative of cryoturbation	0.3 to 0.5

The contact between the loam (silt) and the undulating gravels is cryoturbated and there are ventifacts near the top of the gravel.

A section in the sand and gravel on the south-facing side of the pit [9348 8990] showed:

	Thickness m	Depth m
Gravel fine- to medium-grained, with a silty clayey sand matrix	0.15	0.15
Sand, fine- to coarse-grained, with thin silty clay laminae in horizontally bedded units	0.30 to 0.50	0.65
Sand, fine- to coarse-grained, in cross-stratified sets up to 0.3 m thick, with fine- to medium-grained gravel interbeds; the base is gradational; the foresets of sand units rest directly on the irregular upper surface of the gravel	0.75 to 0.80	1.45
Sand, medium- to coarse-grained, passing downwards into fine- to coarse-grained gravel with a sandy matrix in bedded		

	Thickness m	Depth m
units: cross-stratified sets and cosets up to 0.1 m thick are present in both the sand and gravel	2.50	3.95
Gravel, fine- to medium-grained and fine- to coarse-grained sand in cross-stratified sets	0.50	4.45

The underlying London Clay is not exposed but was proved by augering.

The Star Lane Brickpit [937 873] near Great Wakering worked the terrace loam. The north face of the pit [9335 8749 to 9391 8751] was extensively overgrown and degraded in 1972. Exposures were of weathered yellow-brown sandy clayey silt. The southern and most recently worked face provided a good section [9337 8700 to 9357 8699].

	Thickness m	Depth m
Silt, sandy, yellow-brown (non-calcareous)	1.00 to 1.50	1.50
Silt, sandy, yellow-brown with calcareous cemented root tubules	3.00	4.50
Silt, sandy, clayey, grey-green	0.30	4.80

Mill Head Brickfield is another extensive area of old workings in the terrace loam. The only exposures seen were in the degraded faces. The non-calcareous/calcareous interface was discernible locally [9399 8823, 9429 8832], but elsewhere the loams were apparently uniform. A section [9399 8823] showed:

	Thickness m	Depth m
Silt, sandy, clayey, yellow-brown (non-calcareous)	0.70	0.70
Silt, sandy, brown ('parabraunerde' weathering horizon)	0.30 to 0.45	1.15
Silt, sandy, orange-brown with calcareous-cemented root tubules	1.00	2.15

Another section [9429 8832] also showed the brown weathering horizon to lie between non-calcareous and calcareous silts.

Exposures in the extensive North Shoebury Brickpit are in First Terrace loam; the underlying sand and gravel was proved by augering. Typical sections [9324 8650; 9330 8642] showed:

	Thickness m	Depth m
Soil		0.3	0.3
Silt, sandy, brown to yellow-brown, containing some organic debris and a few flint pebbles, rare horizontal laminations; strong vertical fissuring	1.4 to 1.5	1.8
Silt, sandy, yellow-brown to pale yellow-brown, with anastomosing network of calcareous cemented root-holes forming pale tubules; poor vertical fissuring; flint pebbles rare in the upper parts but more common at depth	1.8	3.6
		(base not seen)

The non-calcareous/calcareous interface produces a marked colour change on weathered sections. The boundary is subhorizontal parallel to the surface. MRH

BRICKEARTH; INCLUDING HEAD BRICKEARTH

Deposits mapped as 'brickearth' differ only in their topographical position from those mapped as River Loam in adjacent areas (p.33). Typically they mantle gentle hillslopes with an easterly and northerly aspect, suggesting an aeolian, loessic origin (p.27); in contrast the overbank River Loam, tends to form the uppermost layer of ancient fluviatile flood-plain deposits, unless indeed it represents an aeolian deposit on the flat top of a terrace (p.33). The brickearth is generally a structureless pale, yellowish brown silt with scattered rootlet traces; locally faint laminations are present and there are a few gravelly wisps. The latter may indicate that the aeolian deposits have suffered some solifluction, and provides justification for the term 'Head Brickearth' to cover cases where the Brickearth has been involved in mass movement. Good exposures are required in order to separate 'Brickearth' from 'Head Brickearth' and, as most of the spreads have been delimited by hand augering, it proved generally impossible to make the distinction. Solifluction appears to have been active at the margins of the Brickearth, for this passes laterally, with admixture of sandy, clayey and stony material, into typical Head. In these cases the limits of the brickearth are ill-defined.

Silt is the predominant component of the Brickearth, with up to 20 per cent of clay, and normally less than 20 per cent of gravel and sand. Columnar jointing, which generally extends down to 2.0 m below ground level, and pale blue or bluish green veins resulting from reduction around rootlet traces, are characteristic. Similar deposits in Kent are known to be metastable and are liable to collapse and flow when wet (Institute of Geological Sciences, 1976, p.22).

In the Eastwood area [835 890; 839 888], Brickearth mantles predominantly north- and east-facing London Clay slopes: it comprises up to 10 m of pale brown silt with a few pebbles, and overlies the London Clay up to about 37 m above OD. Faintly laminated silts were exposed in Eastwood Brick Pit [834 899] at up to 30 m above OD (p.39). Grain size analysis by Mr G. Strong of samples from this pit indicated a loessic component in the sediments, because as much as 50 per cent by weight of the particles had diameters of between 4 and $7\frac{1}{2}$phi (6 to 63μ).

At Cherry Orchard Lane [851 894], Brickearth occurs up to about 30 m above OD and appears to be banked against the London Clay. Other deposits occur on east-facing slopes at Hawkwell [852 927] and Stroud Green [855 910]. At Cherry Orchard Lane, the Brickearth, which contains pebbly stringers, passes laterally into river loams which overlie the Crouch Second Terrace gravels (Figure 14). As in the terrace loams, the Brickearth can be divided into an upper non-calcareous and a lower calcareous part. It is not clear whether it is a partly soliflucted loessic deposit banked against the London Clay, as at Eastwood, or the cryoturbated upper silts of the buried channel deposits (p.27), which may themselves also include a loessic component.

Deposits of Head Brickearth are up to 2 m thick north of Bournes Green [905 875]. They consist of stiff brown clayey silts, yellow-brown silty sands, and sandy silts, with coarse-grained sand lenses and scattered flints. They mask the step between the basal benches of the Second and Third Crouch terraces, which are cut in the London Clay. RAE, GWG

Details

Brickearth

Two large spreads of Brickearth mantling east and north-facing slopes, have been mapped, near Eastwood. In the more northerly, a brickpit [8350 8998] at the north end of Eastwood Rise exposed the following section in 1972:

	Thickness m
Silt, pale brown with rare flints	6.7
Silt, pale brown, with faint laminations	1.8

Augering at the base of the pit proved a further 0.5 m of well-laminated pale yellowish brown silt, which rested on 1 m of brown silt with flints. The lowest 1.5 m of strata was waterlogged, and hence probably lay near to the junction with the underlying London Clay.

The more southerly spread has been worked extensively for brickmaking in the past at a number of localities. The best section in 1972, which has since been filled in, was at the former Bellhouse Brickyard [8393 8877] and showed:

	Thickness m	Depth m
Silt, pale brown	up to c.8.5	8.5
Sand, silty, fawnish grey, thin parting of flint pebbles at base	1.2	9.7
Clay, stiff, grey brown (London Clay)	1.5	11.2
		RAE

Head Brickearth

At Rebels Lane, Sutton [9074 8741], a pit showed:

	Thickness m	Depth m
HEAD BRICKEARTH		
Silt, clayey, firm, brown, with angular, coarse sand grains and scattered subangular and subrounded pebbles grading down into silty fine- to coarse-grained sand with scattered gravel; sharp base	0.85	0.85
Clay, silty, firm, mottled brown and grey, with some sand grains and gravel in the top 20 cm; transitional base	1.25	2.10
?BURIED CHANNEL DEPOSITS		
Silt, clayey, soft, mottled brown and grey, with chalky 'race' nodules up to 3 cm diameter; fine to coarse gravel mostly subrounded occurs throughout; below 3.45 m the silt is greenish grey with common thin-shelled molluscs	1.65	3.75
LONDON CLAY		
Clay, silty, stiff, brown, with scattered mica flakes (augered)	0.15	3.90

The beds below 2.1 m are probably part of the fill of the Rochford Channel.   RAE

HEAD

Those deposits which do not fall into a specific genetic division have been grouped as Head. They are variable in composition, and commonly have been derived from local parent material by solifluction and hillwash involving the sludging downslope of water-saturated material. The detailed lithologies within the Head thus reflect the upslope source material. Thus the Head downslope from the Bagshot Beds is a sandy silt, and that downslope from the London Clay is a silty clay. Silt, gravel and sand, all derived mainly from Terrace Deposits, constitute a highly variable amount of the Head. Sandy and gravelly lenses and pipes are common, and many are contorted. Gravels incorporated in Head at heights above the main terrace sequence have probably been derived from previously more extensive spreads of high-level terraces and Glacial Sand and Gravel of which only small outliers now remain. Naturally, the variety of materials within the Head leads to a considerable variation in its geotechnical properties, both laterally and vertically, over short distances.

Head deposits commonly occur as sheets covering gently sloping ground or as infillings of hollows and valleys. Thicknesses vary from up to 0.5 m in the hillwash which veneers much of the district, to the 3.5 m of Head found in a borehole near Sutton [8957 8840]. Only deposits thicker than 1.0 to 1.5 m are shown on the 1:50 000 Sheet.

In the catchment of the River Crouch, roughly south-west of a line between Wickford and Thundersley, the remnants of a spread of Head about 2.0 m thick occur at 23 to 33 m above OD, overlying London Clay. The surface of these deposits dips very gently northwards. Pits at Bowers Gifford [761 888] within the higher part of the spread showed a veneer of sandy gravelly clays only 0.5 m thick overlying London Clay. Head in the tributary valleys of the River Crouch comprises up to 1.5 m of sandy clays and sandy silt.

Head caps the uplands around Thundersley, Rayleigh and Hockley, where it consists mainly of a soft, grey-mottled, silty clay with a variable pebble and sand content overlying Bagshot Beds and Claygate Beds. The clay is probably derived from Tertiary clays.

There are spreads of soliflucted material derived from Bagshot Beds, Claygate Beds, London Clay and terrace deposits within the valleys that drain south-eastwards from the Rayleigh and Thundersley areas and on the flanks of London Clay ridges near Canewdon, Stambridge and Bournes Green. The lithology varies from silty clay to silty sand, with a variable gravel content that is commonly concentrated at specific levels, particularly at the base of the deposit. The solifluction deposits in the valleys that run eastwards from the upland around Rayleigh commonly include the products of two or more phases of deposition. Probably these comprise separate solifluction lobes of varying lithology which were deposited during the same periglacial episode; less probably they represent the products of separate periglacial episodes.

Head deposits also cover back-features to terraces near Stambridge, Ashingdon and Bournes Green. Gravel, which constitutes up to 30 per cent of the bulk, is embedded in a clayey matrix with coarse-grained sand. A maximum of 3 m was proved in a borehole [8884 9203] west of Great Stambridge. In places, Head has sludged from the features and obscured the Terrace Deposits beneath, thus establishing that the Crouch First Terrace pre-dated the last major period of Head formation.

Details

Upper part of the Roach Valley

Sections in a number of trial pits have been recorded in extensive spreads of Head which occur in the upper reaches of the Roach valley west of Cherry Orchard. A brown clayey silt, which is commonly calcareous in its lower part, overlies a variable thickness of gravel with a sandy clayey matrix.

Thundersley area

Trial pits dug near West Wood, Daws Heath, showed variations in the composition of the Head veneer which masks Bagshot Beds, Claygate Beds and London Clay. One pit [8078 8821] showed two solifluction phases. The upper unit was derived predominantly from the Bagshot Beds:

	Thickness m
HEAD	
Sand, fine-grained, compact, yellow-orange, unstratified, with fine-, medium-and coarse-grained gravel. Small pipes extend down 2.5 cm into the bed beneath	0.65 to 0.90
Clay, stiff red-brown, containing sand grains, concentrated in patches, with scattered pebbles which are mostly rounded black Tertiary flints; erosive base on stratified Claygate Beds below	0 to 1.2
CLAYGATE BEDS	seen 2.0

RDL

CHAPTER 7

Quaternary: Flandrian

MARINE AND ESTUARINE ALLUVIUM AND UNDERLYING BURIED CHANNELS

Extensive spreads of alluvial deposits are present in the Canvey Island and Foulness Island areas and extend up the main river valleys of the district. Most of these sediments have been deposited since the beginning of the Flandrian (Recent), when sea-level stood at around 45 m below OD (D'Olier, 1972), and since when the sea has progressively inundated the low-lying areas. In the absence of reliable faunal evidence, or diagnostic sedimentary structures, it is difficult to separate ancient freshwater alluvium from ancient marine and estuarine alluvium, particularly where strips of alluvium border a river that flows seawards into an estuary. Some arbitrary decisions were, therefore, necessary in classifying the deposits on the 1:50 000 Sheet and producing the following account.

Beneath the alluvial deposits, boreholes and geophysical surveys have proved a complex buried topography which was principally developed by fluvial erosion in response to Pleistocene sea-level fluctuations. Sea-levels during the Pleistocene are estimated to have fallen by 100 m to 130 m (D'Olier, 1972). Melt water from ice sheets during glacial phases, and rivers during glacial and inter-glacial phases, cut distributary patterns to form what are now buried channels, and then filled the channels with fluvio-glacial and fluvial deposits. For convenience, these are referred to as the 'offshore' buried channels to distinguish them from the earlier 'onshore' buried channels of Rochford, Shoeburyness and Burnham-on-Crouch (Table 7).

During the Flandrian, the marine inundation progressively modified a pre-existing topography and partially reworked the earlier Quaternary sediments. This produced complex and variable sequences within the Quaternary deposits. For convenience the entire Pleistocene sequence within the areas covered by marine alluvium are described in this chapter; most are Flandrian, but some are Devensian: direct comparisons are drawn in some instances with contemporaneous deposits which occur elsewhere ('onshore').

The two major alluvial areas of Foulness Island and Canvey Island are described separately below, mainly for geographical reasons. There is a gap of about 10 km in the outcrop of these materials on land between Leigh-on-Sea and Shoeburyness. The alluvium of the two areas reflects somewhat different environments of deposition; the alluvium of Canvey Island has a less marine aspect than that of the Foulness-Maplin area because it was mostly deposited in an inner estuarine environment.   RDL, MRH, BWC

CANVEY ISLAND AND PITSEA

The Shell Haven-Canvey Island-Pitsea area lies in the Thames Estuary where the width of the floodplain is some 10 km; this narrows rapidly to 3.5 km at East Tilbury, 8 km upstream. This change of width is due to the fact that, west of a line from East Tilbury to Cliffe, the river was confined within a channel cut into Chalk, whereas east of that line the channel was able to spread laterally in the softer Tertiary beds. The present river margins of the floodplain are the result of reclamation which took place between the 13th and 15th centuries (Linder, 1940; Evans, 1954).

Sequence

Figure 17 shows a contoured subdrift surface, which is cut in London Clay and Lower London Tertiary beds. The lowest point on this surface is 33.9 m below OD at the eastern end of Canvey Island [8212 8330]. The eastern half of the northern margin of the channel in which the alluvial deposits lie, from Hadleigh to South Benfleet, slopes southwards at a gradient of between 1 in 10 and 1 in 25 to reach a depth of 20 m below OD. The remainder of the northern margin and the western margin, from South Benfleet via Bowers Gifford to Fobbing and Corringham, shows a lower but still marked slope to the south-east and east of between 1 in 60 and 1 in 80, down to a depth of 15 m below OD. The central part of the area shows low gradients, between 1 in 150 and 1 in 200, from a depth of 15 to 20 m below OD. Rock-head in the southern part of the area, around Fobbing Level and the south-west part of Canvey Island, tends to undulate, with a number of enclosed elongated 'lows' below 25 m below OD, which trend east-north-eastwards. Over the eastern part of Canvey Island the channel floor falls gently to an enclosed elongated 'low', below 30 m below OD, which trends north-eastwards.

The marine and estuarine alluvium of this area comprises clays, silty clays, silts, thin peats and silty sands; these are underlain almost everywhere by sandy gravels that fill 'offshore' buried channels and may be fluvial. The whole succession reaches a maximum thickness of about 35 m at the eastern end of Canvey Island. These sediments may be divided into four broad lithological units as follows:

Unit 4: Upper silty clays and clayey silts
Unit 3: Silty sands
Unit 2: Lower silty clays
Unit 1: Sandy gravels

There is much variation in the thickness and lateral extent of the individual units (see Figure 18). Over the greater part of the area the silty clays and silts of Unit 4 occur at ground surface; small outcrops of the sands of Unit 3 occur near the northern margin and on the eastern end of Canvey Island. Unit 1 is thought to represent Middle Devensian fluvial sediments and Unit 2 to 4 Flandrian intertidal and marsh sediments (Table 4).

UNIT 1: SANDY GRAVELS   This unit forms the basal part of

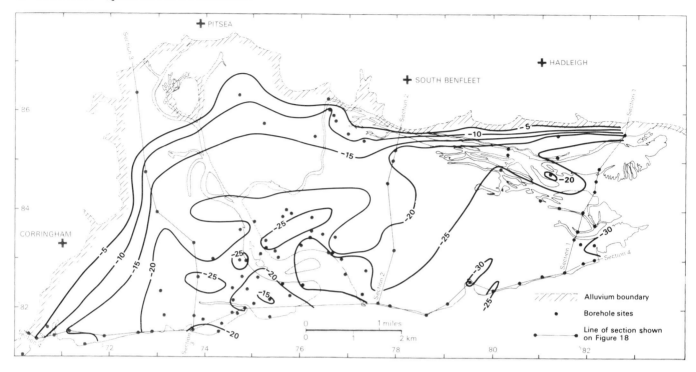

Figure 17 Canvey Island area: contours, in metres relative to OD, on the suballuvium surface

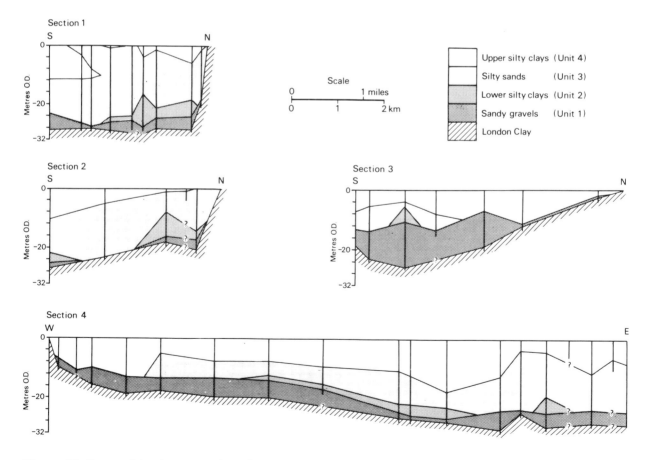

Figure 18 Canvey Island area: sections through alluvial deposits, showing Units 1 to 4

the alluvial succession throughout most of the area and rests on London Clay or Lower London Tertiary beds. It comprises medium dense to very dense, sub-angular to rounded, well graded flint gravel with some medium- to coarse-grained sand. The top 1.5 to 2 m commonly has a silty clay matrix, and there are scattered pockets of laminated silty clay up to 0.5 m thick within the body of the gravel. Isopachytes of Unit 1 gravels are shown in Figure 19.

More than 15 m of gravel are present near Coryton [7384 8265] within an ENE-trending elongate area of high values. Associated with this elongate area at the western end of Canvey Island there are irregular and abrupt variations of thickness. In the central and eastern parts of Canvey Island changes of thickness are less marked, the maximum value being 6.7 m at the mouth of Thorny Creek [7956 8254].

There are two substantial areas where no gravel is present, and Unit 2 or Unit 3 rests directly on bedrock. The largest of these is at Bowers Marshes in the north of the area; the other around Southwick Farm in the middle of Canvey Island. A smaller area lacking Unit 1 material occurs at Hadleigh Marsh at the north-east corner of the area (see Figure 19).

Unit 1 is thought to be the fluvial fill of a 'first' buried channel, which was cut in Middle Devensian times and filled in latest, Devensian times during an interstadial rise in sea level (Table 4).

UNITS 2 AND 4: LOWER AND UPPER SILTY CLAYS AND SILTS   The sediments of these two units are very similar and, where they are not separated by the sands of Unit 3, they cannot be distinguished with confidence from one another. Both units consist of soft to very soft, grey to greyish black, silty clays and silts, with scattered brown and black organic inclusions, plant debris and shell

Table 4   The late Quaternary chronology of the Canvey Island area and outer reaches of the Thames

Event	Sea level (m OD) [1]	Date in years $\times 10^3$ BP [2]	Quaternary stage [3]
Second Thames buried channel filled (Units 2 to 4)	+2	6–7	Flandrian
		10	
Second Thames buried channel eroded	–30	17	Late Devensian
First Thames buried channel filled (Unit 1)	–10	25	
First Thames buried channel eroded	–70	27	Middle Devensian
		36	

1 Zeuner, 1959 and Evans, 1971
2 Evans, 1971
3 Mitchell and others, 1973

fragments. The clays are laminated in places with silt or fine-grained sand partings, and there are thin layers of medium gravel with sand and shells, particularly at the base. Where Unit 4 occurs at the surface, the top 1 to 3 m is firm, mottled grey-brown, and slightly fissured.

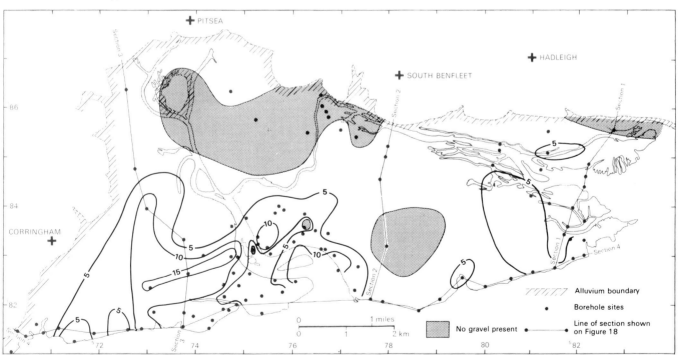

Figure 19 Canvey Island area: isopachyte map of Unit 1 Alluvium (gravels); contours at 5 m intervals

The surface on which Unit 2 rests is contoured in Figure 20. Isopachytes of Unit 2 are shown in Figure 21 for those areas where it can be differentiated from Unit 4 by the intervening sands of Unit 3. The maximum thickness of Unit 2 is 17.7 m and occurs on the north-east side of Canvey Island [8012 8504]. There are two large areas, and several smaller ones, where silty clays are not present beneath Unit 3. The two main areas occupy, firstly, the central and eastern part of Canvey Island and, secondly, parts of Fobbing and Bowers Marshes and the western end of Canvey Island. In the eastern half of the area, Unit 2 thickens relatively abruptly with the contours running parallel to the margins of those areas where the unit is absent. A linear hollow trends north-eastwards through the eastern end of Canvey Island and a bank trending north-eastwards in the central part of Canvey Island veers east-north-eastwards over Hadleigh Marshes. In the southern and western parts of the area changes of thickness are not abrupt; they show no marked trend, nor do they exceed 5 m.

Isopachytes of the silty clays of Unit 4 can be derived from Figure 22 which shows the contoured base of Unit 3. The maximum thickness present is 17.7 m on the south side of Canvey Island [7860 8198], west of Deadmans's Point, where a relatively abrupt increase in thickness results from the presence of an elongated body of clay trending north-eastwards. In the central and northern parts of the island thicknesses decrease gradually to zero over Benfleet Creek along a feature trending a little north of eastwards. At the eastern end of Canvey Island and at Corringham Marshes in the west of the area, changes of thickness are irregular and lacking in orientation. Where Unit 3 is absent, i.e. at Fobbing and Corringham Marshes, isopachytes of the combined thickness of Units 2 and 4 are shown in Figure 21. This shows a regular riverward thickening from the alluvium

boundary. The maximum thickness present is 18.3 m on the south side of Hadleigh Marsh [8031 8522]. The regularity of this trend is disturbed over the western side of Canvey Island by an elongated area trending northwards where there is a much lower rate of increase.

Thin peat beds occur quite commonly in the sediments of these two units, especially in the western part of the area. Two types of peat are present; a compact, amorphous silty or clayey peat and a fibrous woody peat with scattered pockets of silty clay. The peats range in thickness from 0.2 to 0.4 m, with a maximum of 0.7 m. Most of the occurrences appear to fall within the following four broad depth zones: 1.7 m above to 0.7 m below OD (Upper Peat): 2.3 to 6.7 m below OD (Middle Peat): 9.9 to 11.1 m below OD (Lower Peat); and 19.7 to 26.9 m below OD (Basal Peat). The Basal Peat is present only in the central and eastern parts of the area and occurs below the silty sands of Unit 3. These depths equate with those of four similar peats on the Kentish side of the estuary. A sample from the Middle Peat, at a depth of 2.8 m below OD at Bowers Marshes [7472 8635] near Pitsea in the northern part of the area has yielded a radiocarbon date of 5176 ± 65 BP (sample SRR 384; IGS EK111). Radiocarbon dates have been determined for all four beds in Kent. Both units 2 and 4 are thought to be upper tidal flat and marsh deposits, laid down during the Flandrian transgression. They fill a 'second' buried channel cut in Late Devensian times. The peat beds formed during still stands of slight regressions within the overall Flandrian rise in sea level, (Table 4).

UNIT 3: SILTY SANDS   This unit comprises medium dense, blue-grey, clayey, silty, fine- to medium-grained sand with scattered shells. Some pockets of laminated silty clay occur within the sands together with scattered small pebbles of

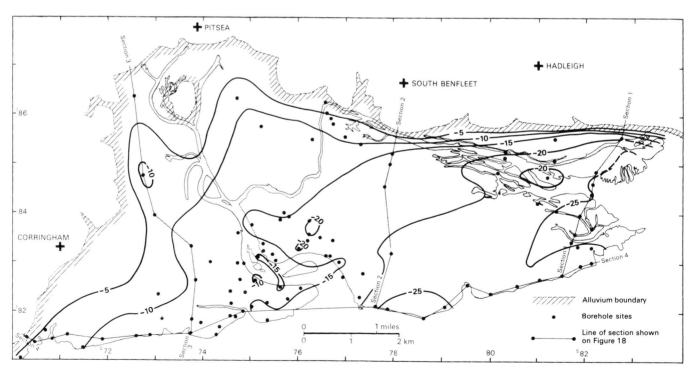

Figure 20 Canvey Island area: contours, in metres relative to OD, on base of Unit 2 Alluvium

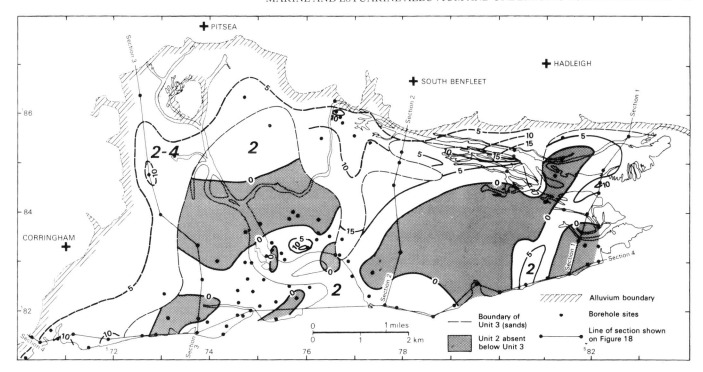

Figure 21 Canvey Island area: isopachyte map of Unit 2 Alluvium (lower silty clays) and Units 2–4 Alluvium (lower and upper silty clays), where Unit 3 Alluvium (sands) is absent; contours at 5 m intervals

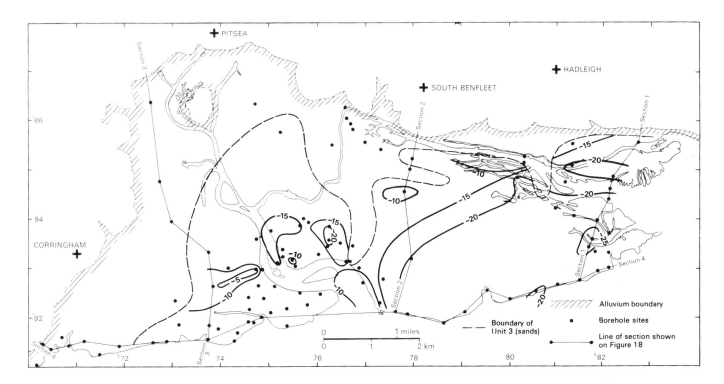

Figure 22 Canvey Island area: contours, in metres relative to OD, on base of Unit 3 Alluvium (sands)

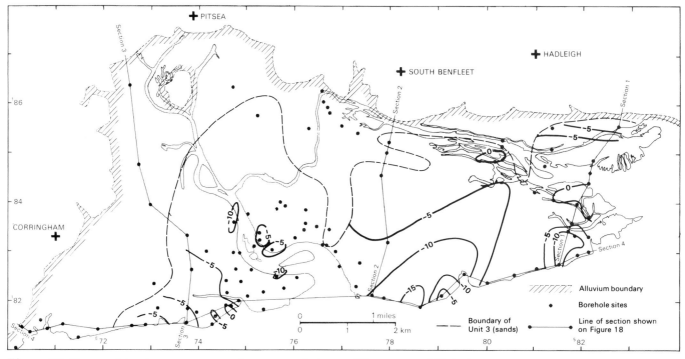

Figure 23 Canvey Island area: contours, in metres relative to OD, on top of Unit 3 Alluvium (sands)

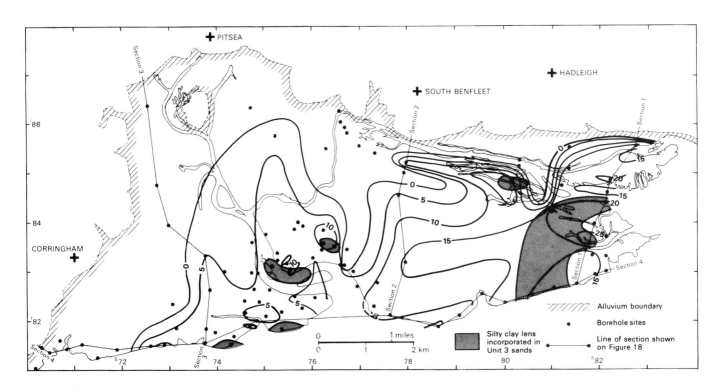

Figure 24 Canvey Island area: isopachyte map of Unit 3 Alluvium (sands); contours at 5 m intervals

chalk and wisps of peat. Localised concentrations of gravel occur in the upper part of the unit. Small surface outcrops of the sands occur near the eastern half of the northern margin of the area, from South Benfleet to Hadleigh Marshes, and also on the western end of Canvey Island.

Figure 22 shows the surface on which Unit 3 rests and Figure 23 shows the top of Unit 3. The main feature of the former surface is a steep-sided, flat-bottomed area which trends east-north-eastwards over the southern and eastern parts of Canvey Island. Figure 24 shows the lobate form of the Unit 3 sand body which thins westwards from a maximum value of 26.9 m at Newlands on the eastern end of Canvey Island [8180 8360]. The areas where this unit is thickest, i.e. in excess of 15 m, are elongated. They trend east-north-eastwards over the western part of the central area of Canvey Island, and north-eastwards over the south-eastern part of the island, coalescing in the Newlands area. The western part of the area shows smaller and less regular changes of thickness. In the west and north-west, at Fobbing and Corringham Marshes, Unit 3 is absent.

The deposits of Unit 3 are thought to be Flandrian lower tidal flat deposits, laid down after a regression at about 8250 years BP.   BWC

FOULNESS ISLAND AREA

The deposits mapped as marine or estuarine alluvium include the muds, clays, silts and sands of the Crouch and Roach valleys, together with those fringing the Thames Estuary near Southend and Shoeburyness, and in the coastal marshlands east of Southend. Locally some freshwater sediments are present, but these are hard to distinguish separately. Thin, impersistent, sandy gravel beds occur at depth overlying the London Clay in the river valleys and in the areas fringing the Thames Estuary. Extensive sandy gravels, some of which are known to infill buried channels and cap erosional benches (see below), overlie the sub-drift surface beneath the wide alluvial plain of the coastal marshlands and the Foulness and Maplin Sands areas. The drift deposits reach a maximum proved thickness of 38.6 m and overlie an irregular buried topography incised into the London Clay (see Figure 13). A series of buried channels, and at least two erosional benches, are known to be present, but the borehole control is not sufficiently adequate to be able to define these in detail. Some of the channels are, however, known to be orientated north–south and others east–west.

The muds, silts and clays crop out at the surface over most of the area but are overlain in the seaward areas of Foulness Island and the Burnham Marshes by sands and silty sands. Neither the basal sand and gravel nor the gravelly sand beds intercalated into the clays and silts come to the surface.

The oldest deposits are the basal sand and gravel, largely of late-Pleistocene age, which may be approximately equivalent to the similar deposits of Unit 1 of the Canvey Island and Pitsea area. They show evidence of reworking during the post-glacial rise in sea-level. The overlying clays, silts and sands are predominantly Flandrian (mainly Boreal to Sub-Atlantic Zones) in age, generally ranging from around 8000 BP to the present day. Deposition in the coastal marshlands was halted by reclamation using enclosures, known as 'inning'. This process continued from the 14th to the 19th century when the last reclamation was made in 1872 on the Dengie Peninsula (Gramolt, 1960; Smith, 1970, p.52).

Sequence

Previously published accounts of the alluvial deposits of the area include those of Greensmith and Tucker (1968; 1969a, b; 1971a, b; 1973) and D'Olier (1972). Evidence for the lithostratigraphical units present and their inter-relationships has been provided both by surface mapping and boreholes. The deposits are grouped into the following four major units:

(iv) Beach Ridge Deposits
(iii) Tidal Flat Sands
(ii) Upper Tidal Flat deposits, including Upper Sand and Gravel
(i) Basal Sand and Gravel

These units are essentially facies assemblages rather than lithostratigraphical divisions. Nevertheless, they tend to occur in the stratigraphical order listed above (see Table 5). Because of their depositional history, all are diachronous and not necessarily horizontally continuous, nor do they always occur in the order given (Figure 25). No firm correlation can be made with the four lithological units of the Canvey Island area.

The sediments are thought to range in age and depositional environment from the late-Pleistocene Basal Sand and Gravel, which is thought to be largely fluvial or glaciofluvial in origin but to have been partly reworked by the transgressing Late-Glacial to Flandrian sea, upwards into the largely Flandrian estuarine, littoral and marsh deposits represented by the upper three units. Marine influences are indicated by North Sea open-shelf foraminifera in the upper part of the succession (p.51).

Basal Sand and Gravel

The Basal Sand and Gravel, directly overlies the London Clay and the Head. In the coastal marshlands and beneath Foulness Island and Maplin Sands, it is an almost continuous, coarse-grained sand and gravel unit, which is generally thickest in the buried channels and thins over the inter-channel areas (Figure 25). Within the Crouch and Roach estuaries and in the Southend and Shoeburyness areas the gravel deposits are thin and impersistent. The unit ranges from coarse-grained gravels with cobbles in a coarse-grained sand matrix, through gravelly fine- to coarse-grained sands with scattered coarse-grained gravel, to sands with gravel and shell material. The constituent clasts, which range from 16 mm to 37.5 mm in diameter, are predominantly angular to subrounded flints with reworked flint pebbles derived from the Tertiary formations. They include a few Lower Greensand cherts, quartzites and vein-quartz pebbles. An alternation of gravelly and sandy units has been observed in boreholes and indicates some form of stratification. Fining-upwards sequences are present in many of the boreholes, and, over all, the deposit grades upwards from a sandy coarse gravel to a gravelly sand. Iron- and calcite-cemented horizons within the basal gravelly unit may possibly have formed under terrestrial weathering conditions.

The thickness of the Basal Sand and Gravel typically varies between 3 m and 7 m, except near buried channels where it is thicker (Figure 25). Thus, for example, 17.07 m were proved in a water well [9995 9648] at East Wick, north of the Crouch Estuary. The top surface of the Basal Sand and Gravel reaches a maximum height of about 3 m below OD and descends to

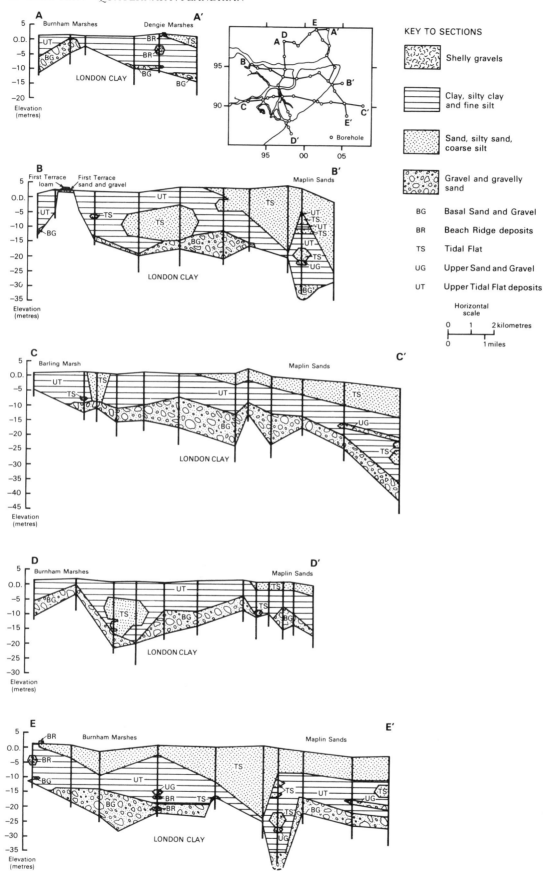

Figure 25 Foulness Island area: sections drawn through the Marine or Estuarine Alluvium succession

Table 5 Lithological units of the marine or estuarine alluvium of the Foulness Island area

Lithological unit	Lithology	Mapped unit	Occurrence
Beach Ridge Deposits	Shell gravels, sandy gravels	Marine or Estuarine Alluvium (Shell Deposits)	Present day beaches and storm beaches, beach ridges, 'chenier ridges' (Greensmith & Tucker, 1969), shell beds in boreholes. Rests discordantly on beds below.
Tidal Flat Sands	Silty sands and sands	Marine or Estuarine Alluvium (Sands)	Reclaimed and present day lower tidal flats. Infilling of past tidal channels. Lower intertidal flat deposits proved in boreholes interdigitate with Upper Tidal Flat deposits.
Upper Tidal Flat Deposits	Muds, clays and silts	Marine or Estuarine Alluvium (Undifferentiated or Clay)	Reclaimed and present day upper tidal flats. Dominant sediment type within the area. River valley sequences are almost exclusively clays and silts whereas in the coastal marshland areas clays and silts frequently comprise up to at least 75% of the deposited thickness.
Upper Sand and Gravel	Shelly, gravelly, sands	Proved in boreholes only	Gravel units intercalated within the Upper Tidal Flat deposits.
Basal Sand and Gravel	Sand and gravel	Proved in boreholes only	Extensive gravel deposits overlying subdrift surface

32 m below OD as the London Clay surface falls away to the south and east. Between the channels underlying Foulness Island it is at around 9 m to 12 m below OD. Within the channels the level of the surface varies, and is controlled by the form of the channel and by subsequent erosion. The upper sandy beds of the Basal Sand and Gravel have been reworked and locally contain marine or estuarine shells and shell debris. The fauna includes *Ostrea sp.*, *?Parvicardium sp.*, and *Corbicula fluminalis*, together with some unidentified gastropods (p.51; Greensmith and Tucker, 1971a, p.307; 1973, p.193). There is a single record of vertebrate remains, namely a worn tooth of the vole *Microtus* cf. *anglicus* (Hinton), which was found in the Basal Sand and Gravel beneath Foulness Island (Greensmith and Tucker, 1973, p.193).

A dominantly fluvial depositional environment is postulated for the Basal Sands and Gravels (Greensmith and Tucker, 1971a, p.307; 1973, p.197) which are thought to have formed during recurrent, high-energy, fluvial and fluvioglacial regimes in Devensian times (p.56). The marine reworking of the upper part of this basal unit is thought to have taken place in Flandrian, or possibly partly in Late-Devensian times (p.51).

Upper Tidal Flat Deposits

The Upper Tidal Flat Deposits lie at the surface throughout the Crouch and Roach estuaries, and in two embayments adjacent to the Thames Estuary near Southend and Shoeburyness, where they commonly form the entire alluvial sequence. In the coastal marshlands they occur at the surface over roughly half the area, extending seawards to the outcrop of the Tidal Flat Sands which overlie and locally interdigitate with them. Throughout this area they are usually underlain by the Basal Sand and Gravel, and they locally contain the Upper Sand and Gravel, Tidal Flat Sands and Beach Ridge Deposits (Figure 25). The deposits are soft to firm, yellow brown to greyish black muds, clays, silts and

silty clays; they vary in total thickness from 2 m to 23 m. Locally they contain sand laminae, thin beds of fine- to medium-grained gravel, shells, shell debris and variable amounts of carbonaceous material, which are indicative of freshwater and hyposaline marshes. Commonly they are extensively bioturbated, but locally they are laminated. The deposits are generally very soft to soft and have a high moisture content. Overconsolidated beds up to 3 m thick (Greensmith and Tucker, 1971a, p.310; 1971b; and below), which are locally associated with peaty clays, occur at depths between 11.5 m and 33 m below OD beneath Foulness Island; peat beds are rare. The topmost 0.5 m to 1.5 m of the deposit have usually been consolidated by desiccation. Locally this desiccated layer thickens to 2.5 m. A relict surface drainage pattern of meandering channels is detectable on much of the marshland.

In the Crouch valley from Battlesbridge to Bridgemarsh Island, the deposits are soft silty clays and clayey silts. Sandy and fine-grained gravelly silts are locally present where the Upper Tidal Flat Deposits overlie the Basal Sand and Gravel. The silt and clay sequences are between 5 m and 10 m thick in areas adjacent to the estuary; they thicken downstream to 15 m at Bridgemarsh Island. These beds thin towards the north and south and are only between 1 m and 3 m thick in the fringing reclaimed areas.

The sediments vary in colour from yellow-brown to pale olive-grey and grey black, with bluish patches. Apart from the oxidised surface layers and the overconsolidated beds, the sediments lie in a reducing environment rich in iron sulphides and hydrogen sulphide gas. The sediments oxidise very rapidly on exposure to air. Organic material is present as carbonaceous debris and as carbonised rootlets. Peaty and organic silts and clays are present and a single peat bed was proved at South Fambridge. It is between 0.2 m and 1.1 m thick, and its base lies between below 1.3 m below and 0.9 m above OD.

In the Roach valley from Rochford to Barling Marsh the deposits are similar to those in the Crouch valley, although much less extensive. Their thickness varies from between 2 m and 5 m at Rochford to as much as 10 m near Barling Marsh.

The Upper Tidal Flat Deposits which fringe the Thames Estuary at Southend-on-Sea [900 850] are 1 m to 5 m thick. They comprise silts, silty clays with sand lenses and organic silty clays, and either rest directly on the London Clay or overlie a thin basal gravel.

At Shoeburyness [932 843], these deposits are 3 m to 5 m thick. At the northern (inshore) end of the deposit, silts and clays are interbedded with sands and gravels and a peaty horizon has been proved. To the south the succession passes into more typical silts and clays with local sandy and gravelly beds, all overlying an impersistent basal sand and gravel.

As the name implies, the sedimentary structures and fauna of the Upper Tidal Flat deposits closely resemble those of modern upper tidal flat deposits, as described by Reineck (1964; 1972) and Reineck and Singh (1973, pp.358–363). The peaty and carbonaceous muds and beds with rootlets thought to have been deposited in supratidal marshes and the overconsolidated beds represent periods of emergence and desiccation. Four still-stands or slight regressions are indicated by these horizons of vegetation colonisation and by horizons of overconsolidation (see below). Throughout the area, overconsolidated beds occur within normally consolidated successions. They can be distinguished by their lower moisture content and liquid limit, by their higher values of apparent cohesion, and by containing layers with higher friction resistance (within silts and clays) in Dutch Cone penetrometer tests. Individual overconsolidated beds reach 3 m in thickness, and the aggregate thicknesses of such beds reaches 5 m to 6 m along the axes of some buried channels. These units are considered to have formed by the desiccation of exposed surfaces (Greensmith and Tucker, 1971b; 1973). Underconsolidated beds have also been recognised in Dutch Cone penetrometer tests in many areas. Those in the upper part of the sequence are the result of rapid sedimentation which has prevented the escape of excess pore waters.

The ostracod and foraminiferal fauna of the Upper Tidal Flat deposits have been examined by Mr M. J. Hughes, who reports they they indicate deposition in a intertidal shallow water environment, with periods of freshwater and hyposaline marsh development but no open shelf connections. Salinities were generally 10 to 25 parts per thousand, but with occasional periods of only 3 to 5 parts per thousand. There is no evidence of temperature fluctuation within the successions examined: this suggests that deposition took place in the middle and late Flandrian (see p.52). The sparse shell fauna comprises *Hydrobia ulvae* (Pennant), *Littorina saxatilis* (Olivi), *Planorbis sp.*, *Spirorbis sp.*, *Scrobicularia plana* (da Costa) and *Cerastoderma lamarcki* (Reeve) (Greensmith and Tucker, 1971a, p.309; 1973, p.195). These species are indicative of deposition in the upper zones of intertidal flats and in the supratidal zones.

UPPER SAND AND GRAVEL   The Upper Sand and Gravel occurs within the Upper Tidal Flat deposits, mainly to the south-east of Foulness Island, typically within the fill of buried channels. It is generally finer-grained than the Basal Sand and Gravel and comprises shelly gravelly sands, sandy gravels and coarse-grained sands with scattered pebbles. The gravel composition is similar to that of the basal deposits. Unlike the basal gravels, marine and estuarine shells, or shell fragments, occur throughout. The Upper Sand and Gravel is usually impersistent, and is between 1 m and 4.5 m thick (see Figure 25). Its top surface lies between 12 m and 18 m below OD and is most commonly between 15 m and 17 m below OD. Deeper occurrences are present within buried channels (Figure 25).

Marine or estuarine shells identified include *Ostrea edulis* (Linnaeus), *Nassarius sp.*, and fragmentary *Corbula sp.* and *Cardita sp.*, together with other bivalves and gastropods. The derived freshwater gastropod *Valvata sp.* was recorded in one borehole [013 912] near Rugwood Head (Greensmith and Tucker, 1971a, p.307).

The material of the Upper Sand and Gravel is thought to have been derived from reworking of the Basal Sand and Gravel and to represent either channel lag deposits of subtidal channels or beach deposits (Greensmith and Tucker 1971a, p.307). Like most of the Upper Tidal Flat Deposits it is thought to be of Flandrian age.

OLDER ESTUARINE ALLUVIUM   At two localities north of Canewdon [9075 9563; 917 952], adjacent to the Upper Tidal Flat Deposits of the coastal marshlands, 1.4 m of firm to soft brown and blue-grey mottled silty clay was proved. These deposits are closely comparable to the adjacent Upper Tidal Flat sediments, but have been distinguished from them by their elevation of around 2 m above OD and their location on a shallow bench cut in the London Clay.

Tidal Flat Sands

The Tidal Flat Sands comprise a persistent unit of soft to firm silty sands, which commonly contain shell debris, and finely laminated silty sands and silty clays. They are found at depth in offshore boreholes; their broad surface outcrop lies along the eastern half of Foulness Island, the mainland area around Wakering Stairs, and the eastern part of the Burnham Marshes (Figure 25). They may be distinguished from the Upper Tidal Flat Deposits by the predominance of sand, the presence of lamination and cross-stratification, and their fauna, which shows open-shelf marine influences.

The Tidal Flat Sands may be divided into three facies, which commonly occur in the order listed below:

(iii) Facies C. Sand, fine- to medium-grained, uniform, grey-brown to pale yellow-brown; uniform; bioturbated towards the top; common shells and shell debris.

(ii) Facies B. Sand, fine-grained, silty, interlaminated with silty clay, brownish grey. Individual laminae may be as thin as 1mm; rarely bioturbated.

(i) Facies A. Sand, fine- to medium-grained, silty, pale yellow-brown to light olive-grey; comminuted shell debris, rarely bioturbated.

Facies A and B can alternate, but Facies C almost always occurs at the top of the sequence. Facies A and B may also occur in the fill of buried channels, at depths as low as 32 m below OD (Figure 25b), as well as being intercalated with the Upper Tidal Flat Deposits, generally in beds 0.5 m to 5 m thick (Figure 25). The top of the Upper Tidal Flat deposits commonly lies between 5 m and 10 m below OD.

Facies A is thought to represent the deposits of subtidal channels, Facies B those of typical intertidal, 'mixed-flat' sediments, and Facies C those of intertidal sand flats (cf. Reineck and Singh, 1973, pp.356-371).

Regarded as a whole, the Tidal Flat Sands tend to thicken seawards. They floor the estuaries of the Crouch, Roach and other recent tidal channels such as Havengore. Within the contemporary channels the sands locally reach 17 m in thickness (see Figure 25). Between the channels, the sands are usually 1 to 7 m thick, but locally reach 15 m in the seaward areas. Only in the infilled estuary channels are all three facies present, as for example in the Proto-Roach eastwards from Monckton Barn [995 935] through Churchend to Courtsend and Fisherman's Head [036 935]. Between the infilled estuary channels, only Facies C is present and here usually rests with a sharp, gently undulating erosional base on underlying Upper Tidal Flat Deposits.

Mr Hughes reports that the ostracod and foraminiferal faunas suggest deposition in water ranging in salinity from 10 to 34 parts per thousand. Open-shelf foraminifera are present in low numbers. A shell fauna of articulated valves and comminuted debris is common in Facies A and C and is of intertidal and subtidal aspect (Greensmith and Tucker, 1971a; 1973). It includes *Cerastoderma edule*, *Mytilus edulis* (Linné), *Macoma balthica* (Linné) *Spisula elliptica* (Brown), *Nucula sp.*, *Barnea candida* (Linné). *Scrobicularia plana*, ?*Parvicardium scabrum* (Philippi), *Ostrea edulis* (Linné), hydrobiids (e.g. *Hydrobia ulvae*), *Corbula gibba* (Olivi), *Buccinum sp.*, *Abra alba* (W. Wood), *Mactra sp.* and some others (Greensmith and Tucker, 1971a, p.309; 1973, p.195). Shells derived from the Tidal Flat Sands are the principal components of the Beach Ridge Deposits (see below).

Beach Ridges

A number of Beach Ridges form laterally impersistent degraded mounds of shell gravel, 10 m to 120 m wide in a zone up to 600 m wide on the Burnham Marshes. They rise little above the general ground level as they have been modified by cultivation and weathering. This zone continues with only small breaks, across the Dengie Marshes to St. Peter's on the Wall, near Bradwell-on-Sea. The ridges are shown on the 1:50 000 map as Marine Beach or Tidal Flat deposits (Shell Deposits). Shell accumulations interpreted as ancient beach ridges have been found at depth in some boreholes to the east of Middle Wick and Holliwell Farm on the Burnham Marshes and east of Churchend and Havengore on Foulness Island. The deposits have an erosional basal contact on the Upper Tidal Flat deposits, and are overlain on the seaward side by the Tidal Flat Sands (see Figure 25c).

Accumulations are forming at the present day adjacent to the sea-wall north of Holliwell Point [0315 9815] and as a prominent, offshore, horse-shoe shaped ridge at Foulness Point [052 957]. These contemporary ridges are formed from shells that have been eroded from the tidal flats, washed into the beach zone, and built into ridges by wave and storm action. They migrate steadily landwards with time, until stabilised by salt marsh vegetation. The shell accumulations

have been described in detail by Greensmith and Tucker (1969a, b; 1971a; 1973).

In boreholes, the shell gravel is composed of broken and abraded bivalves, predominantly *Cerastoderma edule*, with fragmented valves of *Scrobicularia plana*, *Mytilus edulis*, *Ostrea edulis*, *Macoma balthica*, *Nassarius sp.*, *Buccinum spp.*, *Littorina spp.*, and *Hydrobia ulvae*, all contained in a subordinate coarse-grained sandy matrix (Greensmith and Tucker, 1969b, p.421; 1971a, p.309; 1973, p.195). The beds vary in thickness from 0.10 m to 2.9 m.

The Beach Ridges of the Burnham and Southminster Marshes are between 0.5 m and 3 m thick, and are composed of shells and shell debris contained in a subordinate matrix of gravelly fine- to coarse-grained sand with, in places, beds of brown silty clay and clayey sand. Valves of *Cerastoderma edule* again comprise 50 to 70 per cent of the deposit, the other common constituents being *Macoma*, *Mactra*, *Mytilus*, *Ostrea*, *Scrobicularia*, *Hydrobia*, *Littorina* and *Natica* (Greensmith and Tucker, 1969b, p.421).

The contemporary shell ridges have a similar fauna, but this differs from the older ones in containing forms introduced into the southern North Sea since 1600 A.D. These are *Mya arenaria* Linné, *Elminus modestus* (Darwin) *Crepidula fornicata* Linné and *Petricola pholadiformis* (La Mercke) (Greensmith and Tucker, 1969b, p.419).

In the area adjacent to Pig's Bay [952 858], variable and impersistent clayey and sandy gravel Beach Ridge Deposits overlie Upper Tidal Flat Deposits. They are between 1 m and 3.8 m thick, and lie just above mean high water level; they have been subject to channelling and erosion.

Greensmith and Tucker (1969a,b; 1971a) have, by analogy with the coast of Louisiana, called these beach ridges 'cheniers', implying that they formed through the erosion and reworking of pre-existing deposits, during a period of reduced fine sediment supply. The Beach Ridges of this area are, essentially, linear mounds constructed from shells, sand and gravel around high water mark by wave action during storm tides and high spring tides (cf. Reineck and Singh, 1973, p.291–293). Thus it seems preferable to use the more general term 'Beach Ridges' rather than 'cheniers', especially as the depositional environment of the Essex coastal marshes is, unlike that of the coast of Louisiana, subject to strong tidal influences and not subject to the periodic building and abandonment of delta lobes.

Chronological evidence

Table 6 lists radiocarbon dates from the Quaternary deposits in the Foulness area. Dates obtained from shell material in Beach Ridge deposits in the Bridgewick area (Sample Nos. SRR362–3) are anomalous; possibly the organic material was derived from older deposits. All the remaining dates are Flandrian. Detailed chronological subdivision is not possible because of the lack of information, although faunal and floral evidence helps build up a chronological framework. A faunal assemblage in situ including *Corbicula fluminalis*, which is known elsewhere in Britain only from the Pleistocene, was found at 18.70 m below OD in tidal flat deposits in a borehole at Great West Wick [9854 9669]. Further occurrences of *Corbicula* and a worn tooth of the vole *Microtus* cf. *anglicus*, which is restricted to the last glaciation, were record-

Table 6 Radiocarbon age determinations for the Foulness Island area

National Grid reference	Lithological unit and sample material	Locality	Sample number	Radiocarbon age in years BP	Sample depth (m OD)
TR 0142 9971	Beach Ridge Deposit (?including derived shells)	Bridgewick, Borehole TR 09 NW 3	SRR 362 (IGS-EK 82)	14544 ± 95	− 2.3
			SRR 363 (IGS-EK 83)	23625 ± 150	− 3.2
TR 050 940	Tidal Flat Sands: *Ostrea* shells	Maplin Sands, Borehole TR 09 SW 1	IGS-C14/140 (St 3797)	5650 ± 240	− 12.0
TR 029 940	Beach Ridge Deposit: broken *Cardium, Ostrea, Nassarius*	White City, Borehole TR 09 SW 3	IGS-C14/136 (St 3801)	4265 ± 100	− 7.5
			IGS-C14/136 (St 3802)	4350 ± 210	
	Upper Tidal Flat Deposit: peat rich sediment		Birm 242	7516 ± 250	− 18.3
TR 018 943	Beach Ridge Deposit: *Cerastoderma* shells	Ridgemarsh, Borehole TR 09 SW 4	Birm 243	3580 ± 75	− 5.5 to − 8.3
				3912 ± 114	
				3936 ± 110	
	Upper Tidal Flat deposit: *Ostrea* shells, unworn		IGS-C14/139 (St 3798)	6620 ± 100	− 13.8
TQ 9377 9476	Upper Tidal Flat Deposit: *Ostrea* shells, unworn	Wallasea Island, Borehole TQ 99 SW 64	IGS-C14/148 (St 4398)	2245 ± 115	
			IGS-C14/148 (St 4400)	2300 ± 115	− 3.40
TM 021 028	Beach Ridge Deposit	Marsh House Farm	IGS-C14/138 (St 3808)	1340 ± 100	
			IGS-C14/138 (St 3807)	1410 ± 100	+ 0.9
TR 016 994	Beach Ridge Deposit	Court Farm	Birm-244	1265 ± 200	c + 1
				1434 ± 110	(0.5 m depth)
TR 012 995	Beach Ridge Deposit	Court Farm	IGS-C14/137 (St 3800)	645 ± 100	
			IGS-C14/137 (St 3799)	800 ± 100	+ 1.4

ed in the Basal Sand and Gravel at 15.5 m to 17.5 m below OD beneath Foulness Island (Greensmith and Tucker 1973, p.193). Deposits of Pleistocene age have also been recorded on faunal evidence from the inter-channel plateaux. They include much of the Basal Sand and Gravel, but may also include some of the Upper Tidal Flat lithologies.

Analysis of pollen from the Newhouse Farm Borehole [0276 9322], (Tooley, 1973), at a depth of 22.5 m below OD, indicates that Flandrian sedimentation in this sequence commenced in or prior to the Atlantic (Zone V11a) at around 7000 BP; sediments of sub-Boreal (Zone V11b) age, around 5000 BP, occur above 8.45 m below OD. The fauna between 20.95 to 22.45 m below OD in this borehole indicates a temperature which appears to be slightly higher than today and suggests that the Flandrian temperature optimum was at about 5500 to 6500 BP. Evidence of human activity in the adjacent land areas is present throughout the upper 5.70 m

of the borehole, as shown by the presence of pollen of weeds that are associated with cultivation, including *Taraxacum*, *Pteridium*, *Plantago* and *Rumex*.

An attempt to synthesise the Quaternary history of the whole of the district is given on pp.55–57.   MRH

FRESHWATER ALLUVIUM

Freshwater alluvium is shown as 'Alluvium' (undifferentiated) on the published 1:50 000 map. Spreads of freshwater flood-plain deposits occur in the lower reaches of major rivers and their tributaries that are now unaffected by tides. These deposits pass seawards imperceptibly into modern Marine or Estuarine Alluvium. The sediments consist dominantly of soft to firm, grey, organic silty clays.

Lenses of coarser gravelly material occur within the sequence, especially where gravel deposits flank the flood plain. A lag gravel of variable thickness is present locally at the base of the alluvium. In the upper 2 m, the alluvial clay weathers to a brown and pale grey mottled colour, and is commonly penetrated by rootlets.

BLOWN SAND

East of Southend pier, a narrow strip of Blown Sand with interbedded shingle and shell detritus flanks the present coastline. These deposits overlie the Upper Tidal Flat deposits, but are now concealed for the most part by the Esplanade and its associated fill. Boreholes have proved up to 4.1 m of made ground and gravelly sand in this area.

Whitaker (1889, p.478) referred to similar deposits overlying estuarine alluvium near the Crow Stone, Chalkwell [860 853], but these are also now obscured by the Esplanade embankment. Trial holes sited in this area [8581 8538] proved 0.8 m of fill overlying 2.1 m of sand and silt, which in turn rest on estuarine alluvium. RDL

LANDSLIPS

In south-east Essex landslips have been mapped on the north-facing and south-facing slopes of the valley occupied by the River Crouch and its tributaries, and also along the cliffline of the River Thames. Many have been detailed by Hutchinson (1965a,b).

Many of the landslips occur in the London Clay, close to the London Clay–Claygate Beds boundary. This junction is commonly marked by springs which emanate from the sandy, more permeable, layers within the Claygate Beds. Most of the landslipped slopes are probably a result of oversteepening caused by streams which have undercut the foot of the slopes, combined with saturation of the clay by spring water (Denness, 1972).

Landslips at Ashingdon [867 938] and Canewdon [900 948], neither of which is shown on the 1:50 000 Geological Sheet, are entirely within London Clay, and the slopes here are regarded as the remnants of a degraded cliff line of the River Crouch. An area of instability on the west side of Bushy Hill [811 986], near Woodham Ferrers, lies entirely within Claygate Beds, where clay seams cause springs to rise upslope from the Claygate Beds–London Clay contact. Near Gusted Hall [8290 9120 and 8485 9039], London Clay and Claygate Beds slopes, which are subject to spring action but have not been oversteepened, have not slipped. Only a solifluction veneer is found in this area.

Landslipping is widespread along the abandoned River Thames cliff line near One Tree Hill [705 865], Vange [718 872], and between South Benfleet and Southend. Slips occur at the base of the Claygate Beds at One Tree Hill, Vange, and between Benfleet and Hadleigh, and within the London Clay around Pitsea [739 879; 748 879] and between Hadleigh and Southend.

Hutchinson (1965a, b) has made an extensive survey of the landslide areas of Essex and South Suffolk, with particular emphasis on the coastal examples. He noted (1967) that the shallower forms of landslip predominate under present climatic conditions, and that the inclinations of slopes can be broadly related to the type of slip involved (shallow rotational, successive or stepped rotational, 'undulations', markedly non-circular shallow rotational, or predominantly translational). He confirmed that the angle of ultimate stability against landsliding for inland slopes in London Clay and Claygate Beds is 8°, i.e. the same as for those coastal cliffs which do not have a mantle of solifluction deposits.

From detailed studies of the Hadleigh Cliff [814 860], Hutchinson and Gostelow (1976) have documented the history of movements of this feature since it was formed by fluvial erosion in Devensian and early Flandrian times. They observed that the cliff has degraded in an episodic manner, controlled largely by climatic variations. Differing modes of degradation by periglacial solifluction, temperate mudsliding and landslipping were recognised and related to the climatic chronology of the Quaternary. Four main periods of degradation were observed; other comparable European examples appear to relate to the first three of these (Late-Glacial, early Atlantic and early Sub-Atlantic). RAE, RDL

Details

Rotational slips extensively affect the southern margins of the Claygate Beds at One Tree Hill and Vange. At the former locality the movements affect the stability of the road-surface [697 863]: bench and rear-scarp features are well seen at the latter.

Extensive slips along the abandoned river-cliff between South Benfleet and Hadleigh affect the lower part of the Claygate Beds and the London Clay. To the east of Round Hill [794 864], however, active slips have produced a rear-scarp feature close to the base of the Bagshot Beds. Along much of the cliff the slopes are undulating, but in some cases the irregularities of the ground surface are barely perceptible, indicating only minor instability.

The Hadleigh cliffs have been described in detail by Hutchinson and Gostelow (1976), who recorded both the stratigraphical and historical aspects of this tract, based on borehole and trench information, radiocarbon dates from organic material associated with the various colluvium deposits, and the history of the deterioration of Hadleigh Castle. They concluded that, during its history, the cliff has had three toes: an initial one at 19 m below OD, a second one during the Flandrian aggradation at about 9 m below OD, and a final one on completion of this aggradation at 3 m above OD. Hutchinson and Gostelow estimated that the average rate of recession of the cliff was about 4 m per thousand years, until the incidence of a late nineteenth-century landslide which effectively more than doubled the rate.

Between Hadleigh and Leigh on Sea, landslipping is currently more active in the eastern part of the London Clay cliffs where paths, steps and roadways require regular remedial work (Hutchinson, 1965a, b). The cliffs at Southend-on-Sea suffered active marine erosion until the mid-nineteenth century when protection works were commenced. Along their eastern part, they are the steepest in the neighbourhood, with slopes that locally approach 30°. Several slips have been recorded over recent years (Hutchinson 1965a, b). RDL

MADE GROUND

There are several types of artificial fill in the Southend district. The railway embankments are mostly composed of material taken from cuttings along the line and nearby bor-row pits, but slag and ash have also been used locally. Earth-banks of London Clay and subsoil have been constructed in the Basildon area and are used mainly for visual screening.

Other areas of made ground include builder's rubble and domestic refuse. In the eastern part of the district, disused gravel pits have been utilised for dumping this material; in the south-west, it has been dumped on marshland near Mucking, Bowers Marshes and Leigh-on-Sea. Domestic refuse from London is currently transported by barge to Mucking and Holehaven Creek for tipping. Some of these tips have been used for toxic waste disposal.

Earthworks associated with land reclamation and sea-defences are generally too small to show on the 1:50 000 geological map. They are generally constructed of materials such as channel dredgings obtained very locally, e.g. from a borrow area immediately behind the sea-wall known locally as the 'Delph Ditch'.

R.D.L.

CHAPTER 8

Summary of Quaternary history

CHRONOLOGY

The summarised chronology of the Quaternary deposits of the Southend and Foulness district is shown in Table 7. Where possible this is linked to the sequence of events proposed by P.Evans (1971) and Mitchell, Penny, Shotton and West (1973) for the Lower Thames. The relative ages of certain deposits have been discussed in previous sections and these are, therefore, only briefly restated here.

The overall chronology is largely dependant on the geometry and spatial relationships of the various terrace deposits, because definitive evidence such as radiometric dates and diagnostic faunas or floras is comparatively rare. One of the difficulties in trying to construct a Quaternary history is that much depends on the partly subjective correlation of river terraces (see below), or on trying to link deposits with reconstructed curves of relative sea level (Figure 26). Any tectonic or isostatic warping that may have taken place

would obviously affect both of these methods of correlation.

The relationship postulated between the various terrace deposits present in the Lower Thames Valley, based in part on the work of Dines, Holmes and Robbie (1954, p.136) and Evans (1971), to those of the Crouch is shown in Table 7. Here Terraces 2, 3 4 of the south-western part of this district are equated with the Crouch Terraces 1, 2 and 3, respectively. This correlation differs from that of Gruhn, Bryan and Moss (1974, p.70) who equated the Swanscombe (33.5 m above OD) aggradation level with a terrace surface at 12 m above OD (Rochford Terrace) at Southend.

The 'Boyn Hill' terrace, which is mainly of Hoxnian age (see for example Evans, 1971, p.292), is represented by the Fourth Terrace in the Fobbing area and Crouch Third Terrace in the area east of Southend. The Crouch Third Terrace postdates the Burnham Buried Channel and, by inference, similar onshore buried channels.

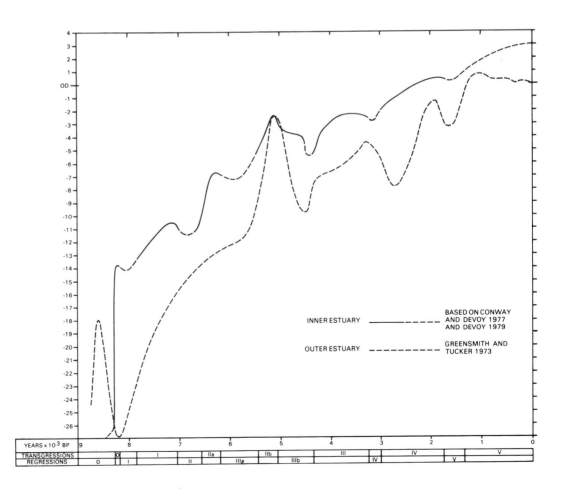

Figure 26 Relative sea-level curves from the inner and outer estuaries of the River Thames

Table 7 Summary of Quaternary chronology

	Canvey Island area and Lower Thames valley	Southend and Foulness area	
Flandrian (= Recent)	Siltation of estuarine area and second buried channel fill (Units 2 to 4)	Siltation of present estuarine areas and east–west channels	
Devensian	Erosion of second buried channel	Main erosion of 'offshore' east–west buried channels	Brickearths & Head Deposits
	First buried channel fill (Unit 1)		
	First Buried Channel eroded		
	'Lower Flood Plain' Terrace 1	Submerged 'bench' gravels (Basal Sand and Gravel, in part)	
		'Offshore' north–south buried channels eroded	
Ipswichian	'Upper Flood Plain' Terrace 2	Crouch Terrace 1	
Wolstonian	'Taplow' Terrace 3	Crouch Terrace 2	
Hoxnian	'Boyn Hill' Terrace 4	Crouch Terrace 3	
		Filling of Rochford, Burnham, ?Shoeburyness 'onshore' buried channels	
Anglian		Crouch Terrace 4	
		Erosion of Rochford, Burnham and ?Shoeburyness 'onshore' buried channels	
		Boulder Clay	
		Glacial Sand and Gravel*	
		Gravels of unknown age†	

* Of northern area
† Possibly older than Anglian

Of the older deposits, the Crouch Fourth Terrace has an irregular distribution profile and occurs mainly in relative close proximity to the Burnham and Rochford buried channels, suggesting a related genesis. The pebble content of these terrace gravels and of the sand and gravel of unknown origin suggests that these deposits may predate the main Anglian glaciation, although the ice striated blocks in the latter deposit present somewhat conflicting evidence unless an earlier glaciation is postulated. Both of these deposits are provisionally regarded as of Anglian age.

Of the younger terrace deposits, the Crouch Second Terrace is apparently poorly preserved in the Southend area and is in places distinguished only with difficulty from adjacent terrace deposits. It is assumed, by inference, to be of Wolstonian age although it could equally be more closely related in time to either of the adjacent terraces. The Crouch First Terrace is equated with the 'Upper Floodplain Terrace' of the Thames, which West, Lambert and Sparks (1964) have suggested is of Ipswichian age though the stratigraphical evidence for this correlation is somewhat ambiguous.

The benches cut into the London Clay and the associated gravel deposits which are present beneath the Flandrian deposits in the Canvey Island and Foulness Island areas are equated with the 'Lower Floodplain Terrace' of the Thames. The north–south oriented buried channels which lie near the bench feature beneath Foulness Island are inferred to be older. The basal sandy gravels of the Quaternary sequence of the River Medway (Cook and Killick, 1924) contain a flora and fauna which reflect cooler climatic conditions than the present day and are thought to be of Middle Devensian age. The basal sands and gravels of the Canvey and Foulness Island areas are also thought to be Devensian. Both sets of gravelly deposits are thought to be dominantly fluvial (pp.41, 49).

The age of the various loams which blanket extensive areas of the Southend district is difficult to gauge. They apparently contain a significant component of loessic material but have been remobilised by cryoturbation and solifluction. Their heights are thus of little value in dating although, where subaqueous bedding is present, parts of deposits may be related to the underlying terrace. In view of their areal extent is is probable that some of these deposits are equivalent in age to similar loams occurring elsewhere in south-east England which have been shown to be early Flandrian. The present of a 'parabraunerde' in the Brickearth (Gruhn and others, 1974) at Cherry Orchard Lane (p.35), however, indicates that some of these deposits may be older, dating back to a warm climatic episode in the Devensian (parabraunerde equivalent age) and beyond.

Most of the marine or estuarine deposits (as opposed to the possibly fluvial basal gravels) of the Foulness and Southend area are known to be Boreal to sub-Atlantic (8900 BP to the present day) in age; no undoubted earliest Flandrian deposits have been recognised. There is little evidence of climatic temperature variation in the faunal assemblages, and the fauna are such as currently inhabit the southern North Sea and Thames Estuary.

The marine or estuarine deposits in the Canvey Island area are also Flandrian (Boreal to Sub-Atlantic) in age (8250 years BP to present day), and were laid down in an estuary progressively inundated by a transgressing sea. Four still-stands, or slight regressions, in the Flandrian marine transgression are indicated by peaty horizons (p.44); two of these are associated with erosional benches on the suballuvial surface.

DEVELOPMENT OF DRAINAGE PATTERN

Following Wooldridge's synthesis of the evolution of the Thames drainage system (1938), it has been generally accepted that the chalky boulder clay ice diverted the Thames from a course trending north-east across Hertfordshire and Essex, to its present alignment (Mitchell and others, 1973, p.48). Gruhn and others (1974, p.70) and Bridgland (1980; 1983) have reaffirmed Whitaker's early hypothesis (1889, p.476) that the Medway and Thames had a common distributary channel across south-east Essex, following this glacial diversion: this argument is based on the presence of significant Lower Greensand cherts in the terrace gravels. The distribution of the Crouch terrace deposits in the Southend–Burnham-on-Crouch area indicates that the successive drainage systems tended to flow in a northerly direction, closely parallel to the onshore buried channels. Buried channels beneath the Flandrian deposits at Foulness also show a northerly alignment, suggesting that this trend was general until Devensian times (see Table 7) when the east–west trend of the present drainage system became dominant. The latter trend was established for the hinterland of the Crouch drainage system in much earlier (Crouch Third Terrace) times.

The origin of the onshore buried channels remains conjectural. Bridgland (1980; 1983) considered their basal gravels to be equivalent to the steeply graded Anglian terraces of the middle Thames but the apparent sinuosity of the Rochford Channel, coupled with its apparent complete siltation and hence abandonment, argues against this. The alternative that the channels have a glacial origin (p.31) remains to be substantiated, although D'Olier (1974, p.256) thought that the enclosed over-deepened channels offshore in the Thames Estuary might be of sub-glacial origin.   RDL

CHAPTER 9

Economic geology

BRICK-CLAYS

The various clays and silts of the district have been utilised extensively in the past for the manufacture of bricks and tiles. London Clay and derived London Clay have been worked at South Woodham Ferrers, at Hullbridge, and east of Fobbing. The Claygate Beds have been dug at Rettendon, One Tree Hill, Hawkesbury Manor, Vange Hall, Down Hall and Hadleigh. Clayey Bagshot Beds have been worked on a small scale east of Stock.

Brickearth and derivatives of the buried channel deposits have been worked at Ballards Gore, Hawkwell, around Eastwood, and at Leigh on Sea. The Cherry Orchard Lane brick-works was active at the time of the survey. River terrace loams were dug at Mill Head (Great Wakering), Thorpe Bay, North Shoebury and Star Lane (Great Wakering). The last-named pit is currently being worked. Where both calcareous and non-calcareous brickearths are worked, as for example at Cherry Orchard Lane, these are normally mixed prior to processing.

SAND AND GRAVEL

Sands and gravels have been worked on varying scales throughout the area (see Chapters 5 and 6). The fine-grained sands of the Lower London Tertiaries and of the Bagshot Beds have also been dug, mainly for domestic purposes. The current sand and gravel resources of the district have been described by Hollyer and Simmons (1978). They comprise mainly the Crouch First Terrace deposits and the sands and gravels associated with the 'onshore' buried channels.   RDL

WATER SUPPLY

The district constitutes part of Hydrometric Area 37. This area was formerly administered by the Essex River Authority, but since April 1974 has been under the control of the Essex Rivers Division of the Anglian Water Authority.

The climate of the district is characterised by a combination of low rainfall and high evaporation. Rainfall distribution follows the relief of the area, varying from less than 535 mm per annum over the eastern coastal areas to more than 585 mm over the slightly elevated areas around Rayleigh and Hanningfield. Seasonal variations in rainfall are minimal, although monthly variations may be more significant. Potential evapotranspiration varies very little across the area, ranging from 530 mm per annum in the north-west to about 540 mm in the south-east. Actual evapotranspiration can vary greatly as a result of soil type and land use differences, but values are, on average, 15 per cent less than the potential, reaching a monthly peak of about 90 mm in June. Soil moisture deficits tend to persist considerably longer than in most parts of Britain, not uncommonly throughout the winter.

The average peak deficit of 125 mm (1956–1965) occurs during July/August and may be long-lived, which tends to delay the seasonal winter increase in runoff and aquifer recharge and creates a maximum water resources depletion around the end of November. Infiltration to the main Chalk aquifer where it is overlain by permeable deposits has been estimated as 100 mm per year, and over areas of less permeable river alluvium as 25 mm per year (Essex River Authority, 1971), with a mean value for the area of the order of 50 mm per annum.

As a result of the low relief and the extensive relatively impermeable cover of London Clay in this part of Essex, the area is drained by numerous small streams which discharge to the sea at Foulness Point via the Rivers Crouch and Roach. In the south several small streams drain a limited area around Pitsea and South Benfleet and discharge into the Thames Estuary at Canvey Island. The only drainage out of the area into other catchments is in the extreme north-west, where the stream on which the Hanningfield pumped-storage reservoir is situated flows northwards to the River Chelmer. The fairly delicate precipitation in evaporation balance of the area causes most of the streams to have low natural base flows, which are unreliable for supply purposes. Flows are measured by Crump Weir at two gauging stations, one on the River Crouch at Wickford [748 934] and one on the Eastwood Brook, a tributary of the River Roach, at Lambeth Road [842 889]. At both, flows of less than 0.02 cubic metres/second are not uncommon. Both the Crouch and the Roach drain urban areas and receive a considerable quantity of effluent which, together with their low natural flows, results in water of variable quality. No direct abstractions for public supply are made from these rivers, but 1.23 million litres per day (Ml/d) is abstracted for agricultural use. Water is also drawn from the Thames for industrial and cooling purposes at the Thames Haven and Canvey Island industrial complexes. Hanningfield Reservoir, in the extreme north-west of the area, stores water pumped from the rivers Chelmer and Blackwater, 12 km to the north; it provides, together with the Langford works, an assessed reliable yield of 114 Ml/d to the Anglian Water Authority for public supply purposes (Essex River Authority, 1971).

Earlier references to the water supply of the area may be found in Whitaker and Thresh (1916), Spens (1955) and various Government departmental reports. A wartime well catalogue (Buchan and others, 1940) records details of wells in the area to that date. For more recent hydrogeological and hydrological data see Essex River Authority, 1971. Discussion of the groundwater chemistry of the area can be found in Ineson and Downing (1963) and Guiver (1972).

The relatively low and unreliable river flows of the area (see above) offer little scope for surface water development, and hence groundwater has been used for supply purposes for many years. With the rapid growth of population and industry in this part of Essex over the last 50 years, water demand has soared, with the result that the groundwater resources of the area are now either developed to their maximum natural capacity or, in some areas, overdeveloped.

Groundwater is available from many of the sand and gravel deposits in the area, but extreme variations in their lithology, saturated thickness and catchment area give rise to highly variable yields. As a result water is abstracted from them only in relatively small quantities for general agriculture and spray irrigation.

The principal aquifers of the area are the Upper Chalk and the Lower London Tertiaries, neither of which has any significant large outcrop with the district. The main recharge to these aquifers is from a small area of Lower London Tertiaries and Upper Chalk exposed immediately to the south-west of the district around Grays, and as a steady leakage from some of the more permeable glacial deposits lying directly on the Chalk and Lower London Tertiaries in northern Essex. Water is also believed to move into the area from recharge areas on the Upper Chalk in Kent: this possibility is surmised from the balance of the water level, abstraction and known recharge statistics. Much of the area is underlain by London Clay which, where thin and weathered in principal valleys and on its outcropping edges, is probably semi-permeable, but which elsewhere is effectively impermeable and a potential aquiclude.

In this part of Essex, the more sandy 'Woolwich type' Lower London Tertiaries are predominant and are generally in good hydraulic continuity with the underlying Upper Chalk. The flow-regime within the Tertiary aquifer is intergranular, whereas that in the Chalk is almost entirely dependent on fissures. Marked variations in permeability occur within the Chalk, and may be related directly to the degree of fissuring. The latter usually decreases with depth, and commonly is highest beneath river valleys, even below a thick Tertiary cover. Transmissivity values beneath valleys are often of the order of 100 to 200 m^2/day whereas elsewhere values are generally less than 100 m^2/day.

The original piezometric surface has been considerably modified by the abstraction of groundwater. In the Tilbury area, immediately to the south-west of the district, groundwater levels are slightly above OD where the Chalk and Lower London Tertiary strata are unconfined. They fall north-eastwards at about 1 in 200 towards Basildon and Southend in the confined area. The average inclinations of the piezometric surface which slopes towards the Southend groundwater depression (see below) from the northern part of Essex is slightly less at 1 in 500. The most noticeable modifications of the original piezometric surface as a result

of pumping, occur in an arc from Basildon through Rayleigh to Southend, and in a small area centred on Burnham-on-Crouch.

In the vicinity of Southend, the piezometric surface now stands at around 80 m below OD, and until recently was continuing to fall at a rate of about 0.7 m a year although still within the London Clay aquiclude. The rate of groundwater extraction clearly exceeds that of groundwater recharge; if this trend were to continue the piezometric surface would be depressed into the Chalk, with a consequent steepening of hydraulic gradient and a possibility of pollution by the intrusion of saline water from the saline front that is present in the Thurrock/Grays area to the south-west. However, at the present time, groundwater levels, which are monitored by 6 observation wells throughout the area, appear to have stabilised, possibly because of reduced abstraction in the main centres of Burnham and Southend.

Over the whole of the district the groundwaters obtained are all of a similar composition and have been classified by the Water Authority as a single hydrochemical group. The water is fairly soft with total hardness values generally less than 80 mg/l $CaCo_3$, no non-carbonate hardness, and with a Ca and Mg content usually less than 10 mg/l. Ph values always exceed 8.0 and are commonly around 8.5. Sulphate ion content is generally greater than 80 mg/l and nitrate content is low, but fluoride content appears to be consistently in excess of 1.5 mg/l and may reach over 4 mg/l. Most of the groundwater of the area has a chloride content of 150 to 350 mg/l and a sodium content of between 250 and 350 mg/l. Exceptionally, in an area around Burnham-on-Crouch both sodium and chloride increase to between 350 and 500 mg/l; this is thought to result from more saline connate water being drawn into the wells from greater depths than elsewhere because of overpumping and the consequent localised lowering of the piezometric surface to 70 m below OD. Representative chemical analyses of groundwaters of the area are shown in Table 8.

Total groundwater abstraction from the combined Lower London Tertiary/Chalk aquifer in this area of south Essex for 1973/74 totalled 2.21 Mm^3 per annum from seventeen major abstraction points. Of this total, over 90 per cent was used for public supply, mainly from boreholes in the Vange and Fobbing areas to the south of Basildon and from an area to the east of Southend-on-Sea. The only large abstraction for industrial purposes is located at the Thames Haven petroleum storage and refinery complex on the Thames Estuary in the south-west, where a group of tidally affected wells pump at individual rates of up to 1.5 ml/d. the relatively small quantities of water used for agricultural purposes are obtained from shallow wells in the minor gravel aquifers in the area. JLF

Table 8   Representative analyses of groundwater from aquifers in the Southend area

Location	Shoeburyness	Burnham (No. 2)	Leighbeck	Prittlewell
National Grid Reference	938 856	947 970	819 834	?
Analyst*	(1)	(2)	(1)	(1)
Date	5.6.69	14.4.71	25.2.69	6.3.69
Classification and thickness	Drift 21.0	Drift 3.6	Drift 31.7	Drift 13.6
of strata (metres)**	LC 100.6	LC 113.1	LC 76.7	LC 110.3
(LC = London Clay	LLT 56.7	LLT 38.0	LLT 48.1	LLT 49.4
LLT = Lower London Tertiaries)	Chalk 35.1	Chalk 44.3	Chalk 35.7	Chalk 92.9
Ca^{2+} mg/l	0.40	0.15	0.50	0.40
Mg^{2+} mg/l	0.41	0.31	0.25	0.33
Na^+ mg/l	15.66	18.27	14.36	15.79
K^+ mg/l		0.14		
Cl^- mg/l	9.14	10.57	8.74	9.53
SO_4^{2-} mg/l	1.91	1.83	1.75	1.60
NO_3^- mg/l	0.02	trace	trace	trace
HCO_3^- mg/l	3.79	5.52	3.31	3.69
Total ionic concentration	31.33	36.79	28.91	31.34
pH	8.4	8.4	8.6	8.4
Total hardness (mg/l $CaCO_3$)	40	23	39	39

* (1)   Pollution Prevention and Fisheries Dept., Essex River Board) now renamed Divisional Scientists Dept.,
  (2)   River Conservators Dept., Essex River Authority) Essex Rivers Division, Anglian Water Authority.

REFERENCES

AKEROYD, A. V. 1972. Archaeological and historical evidence for subsidence in southern Britain. *Phil. Trans. R. Soc. London*, Ser. A, Vol. 272, 151–170.

BADEN-POWELL, D. F. W. 1948. The chalky boulder clays of Norfolk and Suffolk. *Geol. Mag.*, Vol. 85, 279–296.

BAILEY, H. W., GALE, A. S., MORTIMORE, R. N., SWIECICKI, A., and WOOD, C. J. 1983. The Coniacian-Maastrichtian stages of the United Kingdom, with particular reference to southern England. *Newsl. Stratigr.*, Vol. 12, 29–42.

BERDINNER, H. C. 1925. The geology of the Brentwood and Shenfield sections. *Proc. Geol. Assoc.*, Vol. 36, 174–184.

BRIDGLAND, D. 1980. A reappraisal of Pleistocene stratigraphy in north Kent and eastern Essex and new evidence concerning the former courses of the Thames and Medway. *Quaternary Newsl.*, No. 32, 15–24.

— 1983. Eastern Essex in *Field Guide for Annual Meeting. Hoddesdon, 1983.* ROSE, J., (editor). (Quaternary Res. Assoc.)

BRISTOW, C. R. 1971. In *Annual Report for 1970.* INSTITUTE OF GEOLOGICAL SCIENCES. (London: Institute of Geological Sciences).

— 1985. The geology of the country around Chelmsford. *Mem. Geol. Surv. G. B.*

— AND COX, F. C. 1973. The Gipping Till: a reappraisal of East Anglian glacial stratigraphy. *J. Geol. Soc. London*, Vol. 129, 1–37.

— ELLISON, R. A. and WOOD, C. J. 1980. The Claygate Beds of Essex. *Proc. Geol. Assoc.*, Vol. 91, 261–278.

BROTZEN, F. 1936. Foraminifera ausdem Schwedischen, untersten Senon von Eriksdal in Schonen. *Sver. Geol. Undersak*, Ser C. No. 396, 206 pp.

BUCHAN, S., ROBBIE, J. A., HOLMES, S. C. A., EARP, J. R., BUNT, E. F. and MORRIS, L. S. O. 1940. Water supply of south-east England from underground sources (Quarter-inch geological sheets 20 and 24) *Wartime Pamphlet, Geol. Surv.* No. 10.

BURNETT, A. D. and FOOKES, P. G. 1974. A regional engineering geological study of the London Clay in the London and Hampshire basins. *Q. J. Eng. Geol.*, Vol. 7, 257–295.

BURROWS, H. W. and HOLLAND, R. 1897. The foraminifera of the Thanet Beds of Pegwell Bay. *Proc. Geol. Assoc.*, Vol. 15, 19–52.

CASIER, E. 1966. Fauna Ichthyologique du London Clay. *Br. Mus. Nat. Hist.*, 2 Vols. 404–464.

CLAYTON, K. M. 1957. Some aspects of the glacial deposits of Essex. *Proc. Geol. Assoc.*, Vol. 68, 1–21.

— 1960. The landforms of parts of southern Essex. *Trans. Inst. Br. Geogr.*, Vol. 28, 55–74.

CONWAY, B. W. and DEVOY, R. 1977. Cooling Marshes Rochester, Flandrian estuarine deposits of the Lower Thames. *In* Guide to excursion A5, South-east England and Thames Valley. SHEPHARD-THORN, E. R. and WYMER, J. J. (editors). *10th INQUA Congress.*

COOK, W. H. and KILLICK, J. R. 1924. On the discovery of a flint working site of Palaeolithic date in the Medway Valley at Rochester, Kent, with notes on the drift-stages of the Medway. *Proc. Prehist. Soc. East Anglia*, Vol. 4, 133–154.

COOPER, J. 1976. British Tertiary stratigraphical and rock terms, formal and informal, additional to Curry 1958, Lexique Stratigraphique International. *Tertiary Res. Spec. Pap.* No. 1.

CURRY, D. 1958. Great Britain—Palaeogene. *Lexique Stratigraphique International*, Vol. 1, Pt. 3a XII.

— 1965. The Palaeogene Beds of south-east England. Presidential address. *Proc. Geol. Assoc.*, Vol. 76, 151–173.

— 1966. Problems of correlation in the Anglo-Paris-Belgian Basin. *Proc. Geol. Assoc.*, Vol. 77, 437–469.

CURTIS, M. B. and others. 1965. Records of wells in the area of the New Series One-inch (Geological) Dartford (271) sheet. *Water Supply Pap. Geol. Surv. GB., Well Cat. Ser.*

DENNESS, B. 1972. The reservoir principle of mass movement. *Rep. Inst. Geol. Sci.*, No. 72/6, 13 pp.

DEWEY, H. 1912. Report of an excursion to Claygate and Oxshott, Surrey. *Proc. Geol. Assoc.*, Vol. 23, 237–242.

— PRINGLE, J. and CHATWIN, C. P. 1925. Some recent deep borings in the London Basin. Pp. 127–137 in *Summary of Progress for 1924.* GEOLOGICAL SURVEY OF GREAT BRITAIN. (London: Her Majesty's Stationery Office.)

DINES, H. G. and EDMUNDS, M. A. 1925. The geology of the country around Romford. *Mem. Geol. Surv. G. B.*

— HOLMES, S. C. A. and ROBBIE, J. A. 1954. The geology of the country around Chatham. *Mem. Geol. Surv. G. B.*

D'OLIER, B. 1972. Subsidence and sea level rise in the Thames Estuary. *Phil. Trans. R. Soc. London,* Ser. A, Vol. 272, 121–130.

— 1974. Some important stages in the Quaternary development of the Thames Estuary. Centenary de la Société Geologique de la Belgique. L'évolution Quaternaire des bassins fluviaux de la Mer du Nord mériodinale Liège. 253–258.

ELLISON, R. A. 1983. Facies distribution in the Woolwich and Reading Beds of the London Basin. *Proc. Geol. Assoc.*, Vol. 94, 311–319.

ESSEX RIVER AUTHORITY. 1971. First survey of water resources and demands (made under Section 14 of the Water Resources Act 1963).

EVANS, J. H. 1954. Archaeological horizons in the north Kent Marshes. *Archaeol. Cantiana*, Vol. 66, 103–146.

EVANS, P. 1971. Towards a Pleistocene time-scale. Part 2 of the Phanerozoic time-scale: a supplement. *Geol. Soc. London Spec. Rep.*, No. 5, 123–356.

FITCH, F. J., HOOKER, P. J., MILLER, J. A. and BRERETON, N. R. 1978. Glauconite dating of Palaeocene-Eocene rocks from East Kent and the time scale of Palaeogene volcanism in the North Atlantic region. *J. Geol. Soc. London*, Vol. 135, 499–512.

GALLOIS, R. W. and MORTER, A. A. 1982. The stratigraphy of the Gault of East Anglia. *Proc. Geol. Assoc.*, Vol. 93, 351–368.

GRAMOLT, D. W. 1960. The coastal marshland of east Essex between the seventeenth and mid-nineteenth centuries. Unpublished MA thesis, University of London.

GREENSMITH, J. T. and TUCKER, E. V. 1968. Foulness—some geological implications. *Civil Eng. Pub. Works Rev.*, Vol. 63, 525–529.

— — 1969a. Coastline evolution and reclamation in east Essex. *Proc. Geol. Soc. London.*, No. 1969. 313–315.

— — 1969b.   The origin of Holocene shell deposits in the chenier plain facies of Essex (Great Britain).   *Marine Geol.*, Vol. 7, 403–425.

— — 1971a.   The effects of late Pleistocene and Holocene sea level changes in the vicinity of the River Crouch, east Essex. *Proc. Geol. Assoc.,* Vol. 82, 301–322.

— — 1971b.   Overconsolidation in some fine grained sediments; its nature, genesis and value in interpreting the history of certain English Quaternary deposits.   *Geol. Mijnbouwkd.*, Vol. 50, 743–748.

— — 1973.   Holocene transgressions and regressions on the Essex Coast, Outer Thames Estuary.   *Geol. en Mijnbouwkd.*, Vol. 52, 193–202.

GREGORY, J. W. 1915.   The Danbury Gravels.   *Geol. Mag.* Vol. 2, 529–538.

GRUHN, R., BRYAN, A. L. and MOSS, A. J. 1974.   A contribution to Pleistocene chronology of south-east Essex, England. *Quaternary Res.*, Vol. 4, 53–71.

GUIVER, K. 1972.   Chemical characteristics of underground chalk water in Essex and Suffolk.   *Water Treatment and Examination,* Vol. 21, Pt. 1, 30–43.

HARMER, F. W. 1904.   The Great Eastern Glacier.   *Geol. Mag.,* Vol. 41, 509–510.

— 1909.   The Pleistocene Period in the Eastern Counties of England.   Pp. 103–123 in: *Geology in the field, Geol. Assoc. Jubilee Vol.*

HESTER, S. W. 1965.   Stratigraphy and palaeogeography of the Woolwich and Reading Beds.   *Bull. Geol. Surv. G. B.*, No. 23, 117–123.

HOLLYER, S. E. and SIMMONS, M. B. 1978.   The sand and gravel resources of the country around Southend-on-Sea, Essex. Resource sheets TQ 78, 79, 88, 89, 98, 99, TR 08, 09.   *Miner. Assess. Rep. Inst. Geol. Sci.* No. 36.

HULL, E. and WHITAKER, W. 1861.   The geology of parts of Oxfordshire and Berkshire.   *Mem. Geol. Surv. G. B.*

HUTCHINSON, J. N. 1965a.   The stability of cliffs composed of soft rocks, with particular reference to the coasts of south-east England. Ph. D. dissertation, University of Cambridge.

— 1965b.   A survey of the coastal landslides of Essex and south Suffolk.   *Building Research Station Note* EN 36/65. 104 pp.

— 1967.   The free degradation of London Clay cliffs.   *Proc. Geotech. Conf.* (Oslo) Vol. 1, 113–118.

— and GOSTELOW, T. P., 1976.   The development of an abandoned cliff in London Clay at Hadleigh, Essex.   *Phil. Trans. R. Soc. London* A. Vol. 283, 557–604.

INESON, J. and DOWNING, R. A. 1963.   Changes in the chemistry of ground waters of the Chalk passing beneath argillaceous strata.   *Bull. Geol. Surv. G. B.*, No. 20, 176–192.

INSTITUTE OF GEOLOGICAL SCIENCES. 1976.   *Annual report for 1975.* (London: Institute of Geological Sciences.)

KERNEY, M. P. 1971.   Interglacial deposits in Barnfield pit, Swanscombe, and their molluscan fauna.   *J. Geol. Soc. London,* Vol. 127, 69–93.

KING, C. 1981.   The stratigraphy of the London Clay and associated deposits.   *Tertiary Research Special Paper,* No. 6

— 1982.   Comments on 'The nomenclature of the Claygate Beds and Bagshot Beds of London and Essex' (Bristow, 1982) and 'The Claygate Beds of Essex' (Bristow, Ellison and Wood, 1980). *Tertiary Res.*, Vol. 4, 47–52.

KIRBY, R. I. 1974.   Report of a field meeting to Burnham-on-Crouch, Essex.   *Tertiary Times*, Vol. 2, 9–13.

KIRKALDY, J. F. 1933.   The Sandgate Beds of the western Weald.   *Proc. Geol. Assoc.* Vol. 44, 270–311.

KNOX, R. W. O'B. 1984.   Nannoplankton zonation and the Palaeocene/Eocene boundary beds of NW Europe: an indirect correlation by means of volcanic ash layers.   *J. Geol. Soc. London,* Vol. 141, 993–999.

— and ELLISON, R. A. 1979.   A Lower Eocene ash sequence in south-eastern England.   *J. Geol. Soc. London*, Vol. 136, 251–254

— HARLAND, R. and KING C. 1983.   Dinoflagellate cyst analysis of the basal London Clay of southern England. *Newsl. Stratigr.*, Vol. 12, 71–74.

LAKE, R. D., ELLISON, R. A., HOLLYER, S. E. and SIMMONS, M. B. 1977.   Buried channel deposits in the south-east Essex area; their bearing on Pleistocene palaeogeography.   *Rep. Inst. Geol. Sci.*, No. 77/21.

LINDER, E. 1940.   Red-hill mounds of Canvey Island in relation to subsidence in the Thames Estuary.   *Proc. Geol. Assoc.*, Vol. 51, 283–290.

MITCHELL, G. F., PENNY, L. F., SHOTTON, F. W. and WEST, R. G. 1973.   A correlation of Quaternary deposits in the British Isles.   *Geol. Soc. London, Spec. Rep.*, No. 4, 99 pp.

MORTIMER, M. G. 1967.   Some Lower Devonian microfloras from southern Britain.   *Rev. Palaeobotan. Palynol.*, Vol. 1, 95–109.

OWEN, H. G. 1971.   The stratigraphy of the Gault in the Thames estuary, and its bearing on the Mesozoic history of the area.   *Proc. Geol. Assoc.*, Vol. 82, 187–207.

PERRIN, R. M. S., DAVIES, H. and FYSH, M. D. 1974. Distribution of late Pleistocene aeolian deposits in eastern and southern England.   *Nature Phys. Sci. London,* Vol. 248, 320–323.

PIKE, K. and GODWIN, H. 1953 (for 1952).   The interglacial at Clacton-on-Sea, Essex.   *Q. J. Geol. Soc. London*, Vol. 108, 261–272.

PRESTWICH, J. 1850.   On the structure of the strata between the London Clay and the Chalk, etc. Part 1, Basement-bed of the London Clay.   *Q. J. Geol. Soc. London*, Vol. 6, 252–281.

— 1852.   On the structure of the strata between the London Clay and the Chalk, etc. Part 3, the Thanet Sands.   *Q. J. Geol. Soc. London*, Vol. 8, 235–264.

— 1854a.   On the structure of the strata between the London Clay and the Chalk, etc. Part 2, the Woolwich and Reading Series.   *Q. J. Geol. Soc. London*, Vol. 10, 75–170.

— 1854b.   On the thickness of the London Clay, on the relative position of the fossiliferous beds of Sheppey, Highgate, Harwich, Newnham, Bognor, etc and on the probable occurrence of the Bagshot Sands in the Isle of Sheppey.   *Q. J. Geol. Soc. London*, Vol. 10, 401–419.

REINECK, H. E. 1964.   Layered sediments of tidal flats, beaches, and shelf bottoms of the North Sea.   Pp. 191–218 in *Estuaries.*   LAUFF, G. H. (Editor). (Washington: American Association for the Advancement of Science).

— 1972.   Tidal flats. Pp 146–159 in *Recognition of ancient sedimentary environments.* RIGBY, J. K. and HAMBLIN, W. K. (editors) *Soc. Econ. Palaeontol. Mineral. Spec. Pub.* No. 16.

— and SINGH, I. B. 1973.   *Depositional sedimentary environments with reference to terrigenous clastics.* 439 pp. (Berlin: Heidelberg, New York: Springer-Verlag.)

— and WUNDERLICH, F. 1968.   Classification and origin of flaser and lenticular bedding.   *Sedimentology*, Vol. 11, 99–104.

RHYS, G. H. (compiler). 1974.   A proposed standard lithostratigraphic nomenclature for the southern North Sea and an outline structural nomenclature for the whole of the (UK)

North Sea. A report of the joint Oil Industry – Institute of Geological Sciences Committee on North Sea nomenclature. *Rep. Inst. Geol. Sci.*, No. 74/8, 14 pp.

ROSSITER, J. R. 1972.   Sea level observations and their secular variation. *Phil. Trans. R. Soc. London*, Ser. A, Vol. 372, 131 – 140.

SALTER, A. E. 1907.   Excursion to the Laindon Hills, Essex. *Proc. Geol. Assoc.* Vol. 20, 181 – 183.

SCHIMPER, W. P. 1874.   *Traité de paléontologie végétale.* Vol. 3, 896 pp. with an atlas of 110 plates.   (Paris: J. B. BAILLIERE.)

SHEPHARD-THORN, E. R., LAKE, R. D. and ATITULLAH, E. A. 1972.   Basement control of structures in the Mesozoic rocks of the Strait of Dover region, and its reflexion in certain features of the present land and submarine topography. *Phil. Trans. R. Soc. London*, Ser. A, Vol. 272, 99 – 114.

SIMMONS, M. B. 1978.   The sand and gravel resources of the Dengie Peninsula, Essex. Description of 1:25 000 sheet TL90 and parts of sheets TL80, TM00, TQ89, TQ99 and TR09. *Miner. Assess. Rep. Inst. Geol. Sci.*, No. 34.

SMART, J. G. O., SABINE, P. A. and BULLERWELL, W. 1964.   The Geological Survey exploratory borehole at Canvey Island, Essex. *Bull. Geol. Surv. G. B.*, No. 21, 1 – 36.

SMITH, J. R. 1970.   *Foulness. A history of an Essex island parish.* Essex County Council, Chelmsford. Essex Record Office Publications, No. 55.

SPENS, C. H. 1955.   Summary of Essex Water Supply Survey Report. Ministry of Housing and Local Government.

SUTCLIFFE, A. J. 1964.   The mammalian fauna. In 'The Swanscombe Skull.' *Royal Anthropol. Inst. Occas. Pap.* No. 20, 85 – 111.

TERRIS, A. P. and BULLERWELL, W. 1965.   Investigations into the underground structure of southern England. *Adv. Sci.*, Vol. 22, No. 98, 232 – 252.

TOOLEY, M. J. 1973.   Report on the palynological investigations carried out on samples from borehole 1F6A. *Palaeontol. Dep. Rep. Inst. Geol. Sci.*, (Unpublished).

TRUEMAN, A. 1954.   *The coalfields of Great Britain.* 396 pp. (London: Edward Arnold.)

TURNER, C. 1970.   The Middle Pleistocene deposits of Marks Tey, Essex. *Phil. Trans. R. Soc. London*, Ser. B, Vol. 287, 373 – 437.

WARREN, S. H. 1955.   The Clacton (Essex) channel deposits. *Q. J. Geol. Soc. London*, Vol. III, 283 – 307.

WATSON, J. 1980.   Flaws in the continental crust. *Mercian Geologist*, Vol. 8, No. 1, 1 – 10.

WEIR, A. H., CATT, J. A. and MADGETT, P. A. 1971.   Post-glacial soil formation in the loess of Pegwell Bay, Kent (England). *Geoderma*, Vol. 5, 131 – 149.

WEST, R. G. and DONNER, J. J. 1956.   The glaciation of East Anglia and the East Midlands: a differentiation based on stone-orientation measurements of the tills. *Q. J. Geol. Soc. London*, Vol. 112, 69 – 91.

— LAMBERT, C. A. and SPARKS, B. W. 1964.   Interglacial deposits at Ilford, Essex. *Phil. Trans. R. Soc. London*, Ser. B, Vol. 247, 185 – 212.

WHITAKER, W. 1866.   On the 'Lower London Tertiaries' of Kent. *Q. J. Geol. Soc. London*, Vol. 22, 404 – 435.
— 1872.   The geology of the London Basin, Part 1 [part 2 not published]. *Mem. Geol. Surv. G. B.*

WHITAKER, W. 1889.   The geology of London and part of the Thames Valley. Vols 1 and 2. *Mem. Geol. Surv. G. B.*

— and THRESH, J. C. 1916.   The water supply of Essex. *Mem. Geol. Surv. G. B.* 510 pp.

WOOD, S. V. 1868.   On the pebble-beds of Middlesex, Essex and Herts. *Q. J. Geol. Soc. London*, Vol. 24, 464 – 472.

WOOLDRIDGE, S. W. 1923.   The geology of the Rayleigh Hills, Essex. *Proc. Geol. Assoc.*, Vol. 34, 314 – 322.

— 1924.   The Bagshot Beds of Essex. *Proc. Geol. Assoc.*, Vol. 35, 359 – 383.

— 1938.   The glaciation of the London Basin. *Q. J. Geol. Soc. London*, Vol. 94, 627 – 667.

— 1960.   The Pleistocene succession in the London Basin. *Proc. Geol. Assoc.*, Vol. 71, 113 – 129.

— and BERDINNER, H. C. 1922.   Notes on the geology of the Langdon Hills, Essex. *Proc. Geol. Assoc.*, Vol. 33, 320 – 323.

— 1925.   On a section at Rayleigh, Essex showing a transition from London Clay to Bagshot Sand. *Essex Nat.*, Vol. 21, 112 – 118.

— and LINTON, D. L. 1955.   *Structure, surface and drainage in south-east England.* (London: George Phillip.)

WRIGLEY, A. 1924.   Faunal divisions of the London Clay. *Proc. Geol. Assoc.*, Vol. 35, 245 – 259.

ZEUNER, F. E. 1959.   *The Pleistocene Period.* 447 pp. (London: Hutchinson.)

APPENDIX 1

APPENDIX 2

List of 1:10 560-scale maps

The following is a list of the six-inch Natural Grid geological quarter sheets which are included, wholly or in part within the 1:50 000 Southend and Foulness Geological Sheet 258/259, whith the initials of the surveying officers and the date of the survey for each map; the officers are C. R. Bristow, G. W. Green, M. R. Henson and R. D. Lake.

TQ 67 NE (part of)	Muckingford	R. D. L. 1972
TQ 68 NE (part of)	Laindon	R. D. L. 1972
TQ 68 SE (part of)	Stanford le Hope	R. D. L. 1972
TQ 69 NE (part of)	Stock	C. R. B. 1970
TQ 69 SE (part of)	Billericay	C. R. B. 1971
TQ 78 NW	Pitsea	R. D. L. 1972
TQ 78 NE	South Benfleet	R. D. L. 1972
TQ 78 SW	Corringham	R. D. L. 1972
TQ 78 SE	Canvey Island	R. D. L. 1972
TQ 79 NW	Ramsden Heath	C. R. B. 1970
TQ 79 NE	Rettendon	C. R. B. 1970
TQ 79 SW	Wickford	C. R. B. 1971
TQ 79 SE	Rawreth	C. R. B. 1971
TQ 88 NW	Hadleigh	G. W. G. 1972
TQ 88 NE	Southend-on-Sea	G. W. G. 1972
TQ 88 SW	Canvey Island (east)	R. D. L. 1972
TQ 88 SE	Southend Flat	G. W. G. 1972
TQ 89 NW	Stow Maries	C. R. B. 1968
TQ 89 NE	South Fambridge	C. R. B. 1968
TQ 89 SW	Rayleigh	M. R. H. 1971
TQ 89 SE	Rochford	G. W. G. 1971
TQ 98 NW	Barling	M. R. H. 1972
TQ 98 NE	Havengore Island	M. R. H. 1972
TQ 98 SW	Shoeburyness	M. R. H. 1972
TQ 99 NW	Althorne	C. R. B. 1968
TQ 99 NE	Burnham-on-Crouch	C. R. B. 1968
TQ 99 SW	Paglesham	M. R. H. 1971-2
TQ 99 SE	Roach Estuary	M. R. H. 1972
TR 08 NW	Maplin Sands	M. R. H. 1972
TR 09 NW	Dengie Marshes (south)	M. R. H. 1972
TR 09 SW	Foulness Island	M. R. H. 1972

Manuscript copies of these maps are deposited for public reference in the library of the British Geological Survey, Keyworth. They contain more detail than appears on the 1:50 000 map.

Borehole records

Abstracts of lithological logs and lists of fauna collected from selected boreholes

The names of some of the more important borehole records are listed below. Details of boreholes drilled by the former IGS Industrial Minerals Assessment Unit appear in Hollyer and Simmons (1978).

The logs of the boreholes are arranged in order of the 1:10 000 or 1:10 560-scale National Grid quarter sheets in which they occur, and then in numerical order of the BGS Registration Number of the borehole record within each quarter sheet. The unabridged logs of all the boreholes listed may be consulted at the National Geosciences Data Centre in Keyworth, where they, together with the logs of other non-confidential boreholes and wells in the district, may be inspected for a standard fee.

The description of the colours of rock and the accompanying colour code are based on the Rock Color Chart of the Geological Society of America.

Faunal lists for the strata between stated depth intervals are quoted immediately after the lithological description of the strata at the bottom of the specified interval.

Borehole registration No.

TQ 68 NE	
2	Westley Heights Borehole
TQ 68 SE	
33	Stanford-le-Hope (Rainbow Lane) Borehole
TQ 79 NW	
3	South Hanningfield Borehole
TQ 79 NE	
1	Rettendon Borehole
TQ 88 NW	
86	Hadleigh (Sand Pit Hill) Borehole
87	Plumtree Hill, Castle Farm, Hadleigh Borehole
89	Castle Farm, Hadleigh Borehole
TQ 89 SW	
37	Hockley Heights (Gattens) Borehole
TQ 89 SE	
41	Ashingdon Borehole

SHEET TQ 68 NE

2 BGS Westley Heights Borehole [6810 8653]
Surface level + 116.60 m above OD;
Date 1973

	Thickness m	Depth m
BAGSHOT PEBBLE BED		
Pebble bed with coarse-grained sand; black well rounded flint pebbles up to 100 mm diameter with severely chatter-marked surfaces. Thickness approximate	5.00	5.00
Core loss	4.52	9.52
BAGSHOT BEDS		
Clay, very silty, dusky yellow brown (10YR2/2) Laminated. Wisps of very fine-grained sand throughout. Sharp base	0.14	9.66
Clay, silty brown to greyish red (5Y5/6) to 5 YR5/2 and 10R5/2). Very fine-grained sand laminae. Occasional more silty beds up to 20 mm thick, dark yellowish orange (10YR4/6) with iron oxide concentrated in places. Clay wisps throughout (Molluscs, preserved in limonite, including *Ringicula sp.*, *Adeorbis sp.*, and *Spiratella* cf. *tutelina* Curry between 9.55 and 9.80 m)	0.41	10.07
Sand, very fine-grained, moderate brown (10YR5/4); scattered mica flakes	0.16	10.23
Silt, clayey with irregular lenses and wisps of dark grey (N4) clay. Occasional very fine-grained sand partings, medium yellowish brown (10YR5/2), iron stained in part. Increasing very fine-grained sand content to base	0.28	10.51
Sand, very fine-grained, moderate yellowish brown (10YR5/4)	0.54	11.05
Sand, very fine-grained, very silty, colour as above. Well defined base	0.13	11.18
Sand, very fine-grained strongly oxidised and moderate yellowish brown (10YR5/6). Occasional grey (N4) clay beds (up to 10 mm thick) from 11.32 to 11.40 mm. Abundant carbonaceous flecks on a bedding plane at 12.48 m	1.65	12.83
Sand, very fine-grained dark yellowish brown (10YR6/4). Finely interbedded with clay units up to 3 cm thick, dusky red (10R2/2). The clay beds have sharp tops and bases and contain very thin streaks of very fine-grained sand. Irregular thicknesses of clay in any one bed suggest lenticular or flaser-type bedding	0.15	12.98
Sand, very fine-grained, soft, moderate yellowish brown to dark yellow brown (10YR5/4 to 10YR6/4); occasional carbonaceous fragments throughout	4.55	17.53
Sand, very fine-grained, finely interbedded with very silty clay. Variable amounts of clay wisps within the sand beds. Small-scale fining-upwards rhythms up to 70 mm thick; grading upwards from laminated, silty, very fine-grained sand with increasing clay wisps, into silty clay with few very fine-grained sand streaks	0.21	17.74
Sand, very fine-grained, moderate yellowish brown (10YR5/4)	0.26	18.00

	Thickness m	Depth m
CLAYGATE BEDS		
Core loss (in soft sand)	2.55	20.55
Sand, very fine-grained, silty, dark yellowish orange (10YR4/6). Occasional seams of silty clay, pale brown (5YR5/2). Gradational base	0.19	20.74
Sand, very fine-grained (approaching silt grade), pale yellowish brown (10YR6/2). Lenses and streaks containing abundant clay wisps measuring 1 × 5 mm. Small burrows on bedding surface at 20.79 m	0.08	20.82
[Core broken from 20.81 to 20.92 m]		
Silt, compact, dark yellowish brown (10YR4/2). Occasional laminae up to 2 mm thick of very fine-grained sand. Abundant clay wisps throughout	0.06	20.88
Sand, very fine-grained (approaching silt grade), clayey, yellowish brown (10YR6/2), laminated. Pyrite nodule 50 mm in diameter at 21.01 to 21.03 m. Increasing clay content towards base	0.25	21.13
Silt, clayey, dark greenish grey (5GY4/1) coarsening downwards into sand, very fine-grained, clayey, moderate yellowish brown (10YR5/2). Abundant silt streaks in the upper part and thin laminae of very fine-grained sand throughout. Seam of very silty clay, olive grey, from 21.42 to 21.44 m. Evidence of slumping of silty clay seams on very fine grained sand. Pyrite nodule at 21.72 m	0.60	21.73
Sand, very fine-grained, greenish grey. Abundant clay wisps throughout, concentrated in thin bands and occasional clay seams up to 10 mm with sharp tops and bases	0.07	21.80
Sand, very fine-grained; slightly micaceous, scattered glauconite pellets, laminated. [Core very wet and broken, 0.27 m recovered]	1.24	23.04
Clay, very fine-grained sandy. Abundant clay wisps throughout. Lenses and streaks of clayey, very fine grained sand	0.16	23.20
Sand, very fine-grained, silty, olive grey (5Y6/2). Laminated. Sharp erosive base	0.10	23.30
Sand, very fine-grained, silty, dark greenish grey (5GY4/1); glauconitic and slightly micaceous. Many small irregular clasts of very clayey silt. Fining downwards into very clayey silt, olive grey (5Y4/1), finely laminated. Sharp base	0.21	23.51
Silt, brown and iron stained. Sharp base	0.01	23.52
Sand, very fine-grained, glauconitic. Wisps of very clayey silt throughout. Finely laminated coarse silt from 23.79 to 23.81 m. Sharp base	0.34	23.86
Sand, very fine-grained, silty, greenish grey; glauconitic and slightly micaceous. Remnant lamination in lower 0.20 m. Abundant clayey silt clasts throughout, particularly in the upper 0.40 m. Clay wisps and lenses (lenticular lamination) in the basal 0.20 m	0.70	24.56
Silt, clayey, olive grey (5Y3/1). Patches of silty, very fine-grained sand, light olive grey (5Y5/2 to 5Y6/2). Burrows up to 5 mm across filled with glauconitic very fine-grained sand. Scattered small bivalves	0.07	24.63

	Thickness m	Depth m

Sand, very fine-grained, silty, yellowish grey in the top 0.08 m, becoming darker and greenish grey (5GY3/1); glauconitic. Finely cross-stratified from 24.66 to 24.76 m with interbedded burrowed clay silt units up to 10 mm thick; the burrows are infilled with very fine-grained, silty sand. Planar stratification from 24.76 m to base with small scale, cyclic sedimentation, grading from finely laminated, silty, very fine-grained sand through clayey silt to silty, fine grained sand between 24.89 and 24.94 m. The more clayey beds are burrowed with very fine-grained sand infilling. Abundant carbonaceous fragments on bedding and lamination surfaces between 24.85 and 24.87 m. Microfossils in the clayey seams — 0.31 — 24.94

Clay, very silty, olive grey (5Y4/1). Sharp base — 0.01 — 24.95

Sand, very fine-grained, dark greenish grey; glauconitic. Bioturbated with relict primary laminations. Clay wisps, up to 30 mm long, throughout — 0.84 — 25.79

No recovery — 1.04 — 26.83

Silt, clayey, olive grey; slightly micaceous. Bioturbated. Abundant streaks and occasional laminae of buff (5Y6/1) silt and dark greenish grey (5GY4/1), very fine-grained sand often partially reworked. Seams up to 5 mm thick of laminated, very fine-grained sand from 27.65 to 27.88 m. Slump structure in sand at 27.33 m — 1.05 — 27.88

(pyrite moulds of gastropods present between 20.55 and 27.88 m)

No recovery — 2.98 — 30.86

Sand, very fine-grained, silty, greenish grey; slightly micaceous. Bioturbated throughout with many clay wisps and dark grey green streaks of pyritous sand. Becoming increasingly clayey towards base and grading into olive grey, clayey silt with buff silt laminae, probably reworked. *Teredo* in carbonaceous material at 32.25 m. Pyrite developed along bedding planes at 31.88 m — 1.44 — 32.30
(Molluscs, including *Ringicula*, *Pachysyrnola sp.*, *Venericardia* cf. *trinobantium*, and very scarce foraminifera and ostracods, between 30.86 and 32.00 m)

Silt, clayey, olive grey. Bioturbated. Occasional streaks of dark grey green, pyritous, very fine-grained sand. Sharp undulating base — 1.16 — 33.46

Silt (approaching very fine-grained sand grade), olive. Bioturbated with relict primary lamination. Occasional pyritous very fine-grained sand patches. Glauconitic from 33.94 m to base. Gradational base — 0.77 — 34.23
(Molluscs, including *Ringicula turgida* (J. Sowerby), *Adeorbis* cf. *lucidus* Cossmann, *Astarte* cf. *filigera* S. V. Wood, *Venericardia trinobantium*, *Nucula sp.*, *Corbula globosa*, *Pachysyrnola sp.*, *Spiratella mercinensis* (Waterlet and Lefèvre), *S. tutelina*, *Camptoceratops prisca*; foraminifera and ostracods are abundant, including 'Cytheridea' sp. nov., between 32.00 and 34.00 m)

	Thickness m	Depth m

Sand very fine-grained, silty, dark greenish grey (5GY4/1) and olive grey (5Y4/1); slightly micaceous. Bioturbated, with burrows infilled with dark greenish grey very fine-grained sand. Increasing silt towards the base (Molluscs including *Corbula globosa*, and common foraminifera and ostracods including *Cytheridea primitia* Haskins, *Paijenborchella sp.*, and *Echinocythereis reticulatissima* Eager, between 34.00 and 35.12 m). — 0.89 — 35.12

LONDON CLAY

Silt, very fine-grained, sandy, olive grey; slightly micaceous. Bioturbated, with patches of very fine-grained sand and occasional relict very fine-grained sand laminae — 0.09 — 35.21

Septarian nodule — 0.10 — 35.31

Silt, very clayey, olive grey. Less silty from 36.22 to 36.43 m and from 37.30 to 37.42 . Scattered dark greenish grey, pyritous, very fine sand laminae and occasional streaks of buff silt. From 39.00 to 39.22 m, buff, very fine-grained sand lenses are frequent, occurring as finely cross-stratified ripple trains, often partially bioturbated; scattered thin shelled unbroken molluscs throughout. Slickensided slip plane at 37.10 m, dipping at 45°. Prominent vertical joint from 39.90 to 40.00 m. Many small gastropods at 37.45 m in a band of clayey silt. Very few lenses and streaks of very fine-grained sand; becoming more frequent from 42.60 m. Gradational base — 8.09 — 43.40
(A varied and well-preserved macrofauna including *Nuculana prisca* (Deshayes), *Euspira glaucinoides* (J. Sowerby), *Adeorbis* cf. *lucidus*, *Pachysyrnola sp.*, *Spiratella mercinensis*, *S. tutelina*. Foraminifera and ostracods are common to abundant including *Cytheridea primitia*, *Echinocythereis reticulatissima*, *Loxoconcha sp.* and *Paijenborchella sp.* between 35.85 and 42.37 m)

Silt, clayey (locally approaches very fine-grained sand grade), greenish grey. Bioturbated. Gradational base — 0.22 — 43.62

Silt, very clayey, olive. Bioturbated. Very fine-grained sand and silt streaks and lenses throughout. Some of the silt streaks contain abundant mica — 0.40 — 44.02

Silt, clayey, very fine-grained sandy. Occasional finely cross-stratified, very fine-grained sand lenses. Homogeneous, with no lenses or laminae between 44.20 and 44.32 m. Lenses increasing in number below 44.32 m, with grey green and olive bioturbated patches of silt. Gradational base — 0.49 — 44.51

Silt, very clayey, olive. Bioturbated. Dark grey-green very fine-grained sand streaks and scattered buff silt laminae — 0.77 — 45.28

Sand, very fine grained, silty; occasional fine-grained sand sized glauconite pellets. Bioturbated with patches of buff silt and dark greenish grey, very fine-grained, pyritous sand and abundant clay wisps. Burrowed base — 0.76 — 46.04

Silt, very clayey, olive grey. Bioturbated. Burrows from bed above filled with greenish grey, very fine-grained sand — 0.06 — 46.10

	Thickness	Depth
	m	m

No recovery 1.90 48.00

Silt, clayey, olive grey, (silt approaches very fine-grained sand grade). Finely laminated, with laminae of micaceous, very fine-grained sand near base — 1.07 49.07

Septarian nodule — 0.27 49.34

Silt, very fine-grained sandy, olive grey. Bioturbated, with patches and clasts of clayey silt — 0.40 49.74

Silt, clayey, olive-grey. Buff silt and dark greenish grey, pyritous, very fine-grained sand streaks throughout. Slip plane dipping at 45° with slickensides from 49.92 to 50.05 m. Increasing very fine-grained sand content towards base — 0.86 50.60

Sand, very fine-grained, silty (sand may be coarse silt grade), greenish grey. Irregular clasts of very clayey silt containing burrows filled with very fine-grained sand — 0.09 50.69

Sand, very fine-grained, silty (as above) olive grey. Bioturbated. Occasional thin buff silt lenses, often finely laminated, from 50.87 m. Becoming gradually finer grained to the base — 0.23 50.92

Silt, very clayey, olive grey. Bioturbated with relict primary lamination in part. Burrows, infilled with buff silt from 51.42 to 51.57 m. From 51.80 m silt and sand streaks present. Becoming more silty towards the base — 0.97 51.89

Sand, very fine-grained, silty (sand may be silt grade), olive grey. Bioturbated, with relict lamination in part. Pyritous, very fine-grained, sand lenses throughout — 0.39 52.28

Silt, clayey, olive. Fairly homogeneous. Bivalves and gastropods common — 0.58 52.86

Silt, clayey, olive. Bioturbated with relict lamination in part. Buff silt streaks and dark greenish grey, very fine-grained sand lenses and streaks. Becoming coarser grained towards the base — 0.21 53.07

Silt, clayey (silt approaching very fine-grained sand grade), olive grey. Occasional thin laminae and streaks of buff, very fine-grained sand and silt. Finely laminated, very fine-grained sand lenses (ripples) from 53.44 to 53.58 m; partially bioturbated. — seen to 0.51 53.58
(Ostracods and foraminifera, molluscs less numerous between 42.64 and 53.58 m. *Pseudohastigerina wilcoxensis* (Cushman & Ponton) abundant at 53.50 m.

SHEET TQ 68 SE

33 BGS Stanford Le Hope (Rainbow Lane) Borehole [6965 8241]
Surface level + 16.14 m OD;
Date 1973

	Thickness	Depth
	m	m

TERRACE 3 DEPOSITS

Gravel, coarse-grained, sandy, predominantly flint, subrounded and subangular. Sand comprises 30 per cent of the bulk (not cored) — 6.40 6.40

LONDON CLAY

Clay, very silty, olive-grey (5Y4/1); micaceous, laminated. Numerous thin silt partings — 0.12 6.52

No recovery 0.43 6.95

Clay, silty, olive-grey (5Y4/1); laminated. Becoming increasingly silty and micaceous towards base — 0.59 7.54
(Rare pyritised diatoms, *Coscinodiscus sp.*, 1 between 6.50 and 7.50 m)

Silt, clayey, brownish grey (5Y4/1); micaceous; laminated. Scattered very thin very fine-grained sand partings. Carbonised plant remains associated with 5 mm diameter pyrite nodules at 7.98 to 8.06 m. Gradational base — 0.52 8.06

Clay, very silty as above. Finely laminated. Occasional very thin very fine-grained sand partings. Carbonaceous debris at 8.26 m — 0.64 8.70

No recovery 1.20 9.90

Silt, clayey, brownish grey (5YR4/1) as above; micaceous. Finely laminated — 0.24 10.14

Silt, clayey, very fine-grained sandy, mottled greenish grey and brownish grey (5YR4/1 and 5GY4/1). Bioturbated, with irregular patches of very fine-grained sand; micaceous — 0.60 10.74

Sand, very fine-grained, silty, greenish grey as above. Scattered bivalves from 10.86 m downwards — 0.20 10.94

Sand, very fine-grained, silty, as above. Clasts of laminated olive grey clay up to 20 mm diameter — 0.80 11.02

Sand, very fine-grained, silty, greenish grey. Broken shell fragments throughout — 0.03 11.05

Sand, very fine-grained, silty, olive-grey (5Y4/1). Bioturbated in the upper 70 mm with irregular patches of very fine-grained sand — 0.94 11.99

Sand, very fine-grained, slightly silty, calcareous from 11.99 to 12.30 m. Horizontal borings infilled with very fine-grained sand from 12.38 to 12.40 m. Bioturbated from 12.75 to 12.97 m. Remnant cross stratification at 12.61 m. Complete *Astarte* shell at 12.47 m with burrow. Wood fragments at 12.44 m. Gradational base — 1.00 12.99

Sand, very fine-grained, dark greenish grey (5GY4/1 to 5GY3/1). Abundant broken bivalve debris and clayey silt clasts up to 20 mm across. Pyritised carbonaceous remains throughout — 0.27 13.26
(Sparse microfauna: *Protelphidium sp.*, 3, Murray & Wright above and shell debris below, between 8.30 and 13.26 m)

	Thickness m	*Depth* m
OLDHAVEN BEDS		
Sand, fine grained, khaki grey, (5Y5/3), scattered broken shells	0.04	13.30
No recovery	6.30	19.60
WOOLWICH BEDS		
Sand, fine-grained, slightly clayey buff; laminated. Scattered disseminated pyrite clusters. Rapid gradation at base	0.31	19.91
Clay, silty, greenish brown (5YR4/1). Laminated, containing burrows up to 10 mm diameter filled with brownish green fine-grained sand. Well developed vertical linear fissure	0.23	20.14
Clay, very silty, grey brown, containing white and black, well rounded, flint pebbles between 15 and 30 mm in diameter. Discontinuity at base	0.06	20.20
Sand, medium-grained, well sorted, subangular to angular, grey-brown. Gradational base	0.02	20.22
Sand, fine-grained, carbonaceous, black (N2)	0.02	20.24
Lignite, laminated, brownish black (5YR2/1)	0.01	20.25
Clay, very silty, black with carbonaceous streaks	0.07	20.32
Lignite, fine grained, sandy, brownish black (5YR2/1). Gradational base	0.09	20.41
Sand, medium-grained, subangular and angular grey-brown (5YR3/1). Structureless	1.25	21.66
Sand, medium-grained, grey-brown (5YR4/1) in the top 0.02 m, grading into pale brownish grey (5YR6/1) medium-grained structureless sand	0.15	21.81
Sand, medium-grained, dark grey-brown in the top 0.04 m, grading into medium- to fine-grained black sand. The black colour is probably an iron or manganese oxide coating to the grains	0.20	22.01
Sandstone, fine-grained with dark brown and greyish brown banding. Cross-stratified channel fill with an erosive base from 22.02 to 22.15 m; reworked clay wisps occur near base of the channel	0.19	22.20
Sandstone, fine-grained, black. Erosive base channelled into bed beneath	0.14	22.34
Sand, fine- to medium-grained, brownish grey. One 5 mm diameter rounded black flint pebble at the top of the bed. Gradational base	0.47	22.81
Sand, fine-grained, speckled medium grey (N4) and bright green. Highly glauconitic	0.67	23.48
No recovery	1.04	24.52
Sand, fine- and medium-grained, slightly clayey and with occasional subangular coarse-grained sand grains, green and pale grey speckled. Scattered round black flint pebbles up to 10 mm diameter at base. Highly glauconitic	0.16	24.68
Sand, fine- and medium-grained, clayey, with occasional round black flint pebbles in the lower part, dark grey (N3) with green glauconitic pellets scattered throughout. Bioturbated; some burrows infilled with fine grained glauconitic, greenish sand. Occasional clay clasts up to 10 mm across	0.81	25.49

	Thickness m	*Depth* m
Sand, very fine-grained, clayey, dark greenish grey. Glauconite pellets and black rounded flint pebbles up to 30 mm long scattered throughout	0.06	25.55
No recovery	3.49	29.04
THANET BEDS		
Sand, very fine-grained, silty, greenish grey (5GY5/1). Highly bioturbated	5.76	34.80
[Core loss from 34.60 to 34.98 m]		
Sand, very fine-grained, silty (greater silt content than above). Bioturbated	2.46	37.26
Sand, very fine-grained, silty, greenish grey as above. Bioturbated, with occasional pyritised burrows	1.34	38.60
Core loss	0.61	39.21
Sand, very fine-grained as above. Gradational base	0.48	39.69
Sand, very fine-grained, silty (greater silt content than above), greenish grey, bioturbated	0.19	39.88
Sand, very fine-grained, silty as above. Gradational base	1.71	41.59
Sand, very fine-grained, very silty, greener than above and slightly micaceous. Gradational base	0.38	41.97
Sand, very fine-grained, silty, greenish grey (5GY5/1). Gradational base	2.05	44.02
Sand, very fine-grained, silty, olive grey (5Y4/1). Bioturbated from 45.16 to 46.54 m. Gradational base	2.52	46.54
Silt, very fine-grained sandy, olive grey (5Y5/1). Bioturbated	2.48	49.02
No recovery	0.58	49.60
Silt, very fine-grained sandy, olive grey (5Y5/1). Bioturbated, with occasional pyritised burrows. Gradational base	1.60	51.20
Silt, very fine-grained sandy, olive-grey as above. Faintly cross-stratified	0.40	51.60
Silt, very fine grained-sandy, greenish grey (5GY5/1). Bioturbated, with burrows, often horizontally orientated, infilled with olive-grey (5Y5/1) very fine-grained sand. Gradational base	2.08	53.68
Silt, greenish grey. Bioturbated, with small patches of very fine-grained sand. Colour changes gradually to olive-grey at around 54.90 m	2.42	56.10
Silt, slightly very fine-grained sandy, olive-grey. Abundant burrows infilled with very fine grained sand, olive-grey in colour. The burrows are mainly horizontally aligned. Disseminated glauconite pellets from 56.22 m, increasing in number towards the base	0.56	56.66
Silt, very fine grained sandy, olive-grey. Scattered glauconite pellets and occasional 20 mm diameter subangular flints and small (5 to 10 mm diameter) black rounded flint pebbles (Pyritised diatoms, *Coscinodiscus sp.* 2, between 46.00 and 56.70 m)	0.09	56.75
No recovery	0.20	56.95

	Thickness m	Depth m
Silt, clayey. Subrounded irregularly shaped flint pebbles ('Bullhead' flints) up to 60 mm diameter, coated with medium-grained sand-sized glauconite pellets. A 0.01 to 0.02 m thick bed of sand, fine-grained, clayey and glauconitic, occurs at the base of the bed. Unconformable on bed beneath (Bullhead Bed)	0.10	57.05

UPPER CHALK
Chalk	seen to 1.51	58.56

SHEET TQ 79 NW

3 BGS South Hanningfield [7413 9715]
Surface level +72.14 m OD;
Date 1973

	Thickness m	Depth m
BAGSHOT BEDS		
Sand, fine-grained, dark yellowish orange (10YR6/6); bioturbated	3.80	3.80
Clay, silty, light brown (5YR6/4) mottled pale blue green (5BG7/2) bedded, finely laminated with some silty sand laminae	0.05	3.85
Sand, fine-grained, dark yellowish orange (10YR6/6)	0.65	4.50
CLAYGATE BEDS		
Sand, fine-grained, silty, clayey, yellow brown (10YR6/4), bedded with subsidiary grey silty clay, laminae	0.80	5.30
Sand, ferruginous, bedded, with silty clay, dark grey (N3), and fine sand laminae	0.30	5.60
Silt, sandy, bedded, with sandy clay yellow brown, (10YR6/4), laminae	1.40	7.00
Silt, clayey, sandy, pale brown (5YR5/2), bedded; silt clay laminae	0.40	7.40
Silt, clayey, sandy (fine), dark greenish grey (5GY4/1); micaceous	1.10	8.50
Clay, very silty, dark greenish grey (5GY4/1); rare septarian nodules, finely laminated, fissured. Fine sand laminae	1.90	10.40
Clay, silty, olive grey (5Y4/1), bedded, finely laminated; fissuring	2.50	12.90
Silt, clayey, sandy, olive-grey (5Y4/1); finely laminated	2.30	15.20
Sand, fine grained, silty, dark greenish grey (5GY4/1)	0.90	16.10
Clay, very silty, olive-grey (5Y4/1); bedded, fissured; sandy, clayey silt laminae, bioturbated	0.90	17.00
Clay, very silty, olive-grey (5Y4/1); bedded, with silt laminae	0.90	17.90
Silt, clayey, olive-grey (5Y4/1); molluscs. Laminated	seen to 2.50	20.40

(The complete fauna from this borehole includes *Adeorbis* cf. *lucidus*, *Aporrhais sp.*, *Cerithiella sp.*, *Euspira sp.*, *Pachysyrnola carinulata* Cossmann, *Ringicula turgida*, *Sinum* cf. *clathratum* (Gmelin),

Camptoceratops prisca, *Spiratella tutelina*, *Anomia sp.*, *Corbula globosa*, ?*Cultellus sp.*, *Nuculana prisca*, *Pitar sp.*, *Cytheridea primitia*, '*Cytheridea*' *sp.*, *nov.*, *Echinocythereis reticulatissima* and *Loxoconcha sp.*)

SHEET TQ 79 NE

1 BGS Rettendon [7702 9606]
Surface level +51.54 m OD;
Date 1973

	Thickness m	Depth m
MADE GROUND	0.90	0.90
CLAYGATE BEDS		
Clay, silty, pale yellowish brown (10YR6/2) with sand lenses and laminae. Laminae of hard pale yellowish orange (10YR8/6) arenite	0.70	1.60
Sand, fine-grained, silty, medium yellow brown (10YR5/6); bedded	2.40	4.00
Clay, silty, pale brown (5YR5/2) grading downwards to clayey, fine sand, pale brown; laminae	1.00	5.00
Silt, clayey, pale brown (5YR5/2) bedded	1.10	6.10
Sand, fine-grained and silty clay, dark greenish grey (5GY4/1) with pyrite nodules. Alternations of silty fine sand and silty clay, laminae, bioturbated, fissuring	2.20	8.30
Sand, fine-grained, silty dark greenish grey (5GY4/1) with silty clay laminae. Bioturbated, fissured	1.30	9.60
Clay, silty, stiff, olive-grey (5Y4/1); bioturbated, fissured. Thin sand, fine grained laminae	seen to 2.70	12.30

(Fauna recorded in borehole: *Rotularia bognoriensis* (Mantell), *Adeorbis* cf. *lucidus*, *Euspira sp.*, *Pachysyrnola carinulata*, *Ringicula turgida*, *Seila sp.*, *Turboella sp.*, *Camptoceratops prisca*, *Cuspidaria* cf. *inflata* (J. Sowerby), *Lentipecten* cf. *corneus* (J. Sowerby), *Nuculana prisca*, *Venericardia trinobantium*, *Cytheridea primitia*, *Echinocythereis reticulatissima*, *Loxoconcha sp.*, *Paijen-borchella sp.*, and *Squalus sp.*

SHEET TQ 88 NW

86 BGS Hadleigh (Sand Pit Hill) Borehole [8002 8654]
Surface level + 70.56 m OD;
Date 1973

	Thickness m	Depth m
TALUS (in part)		
No core, sand, very fine-grained and gravelly in pit to 2.0 m	4.80	4.80
Core loss	2.97	7.77
BAGSHOT BEDS		
Sand, very fine-grained silty, dark yellowish orange (10YR5/6), becoming darker towards base	0.21	7.98
Siltstone, brownish with purple shades in concentrated iron oxides	0.01	7.99
Sand, very fine-grained, silty. Common grey (N4) clay streaks	0.03	8.02
Sand, very fine grained, silty, pale yellow brown (10YR6/6); slightly micaceous	0.33	8.35
Clay, moderate yellowish brown (10YR5/4). Coarsening downwards into clayey silt and then silty, very fine-grained sand at 8.36 m	0.02	8.37
Sand, very fine-grained, silty, yellow-brown as above. Finely interbedded with grey clay, clayey silt and fine sandy silt	0.14	8.51
Silt, very fine-grained sandy, yellowish brown and brown (10YR5/4 and 5YR5/6). Occasional grey clay (N5) streaks up to 1 mm thick	0.33	8.84
Clay, pale yellow-brown to brownish grey (10YR7/2 to 5YR5/1). Finely laminated. Gradational base	0.05	8.89
Sand, very fine-grained, very silty, dark yellowish orange and dark yellow brown (10YR6/6 and 10YR5/2); micaceous	0.19	9.08
Clay, brownish grey (5YR5/1). Finely laminated	0.02	9.10
Sand, very fine-grained, very silty, yellowish brown (10YR5/2) with dark browns and red-browns varying with iron oxide content. Well developed iron pan at base	0.52	9.62
CLAYGATE BEDS		
Silt, very fine-grained sandy, yellowish brown (10YR5/2). Finely laminated. Discontinuous grey clay streaks throughout, between 1 and 5 mm thick	0.08	9.70
Silt, very fine-grained sandy, pale yellowish brown (10YR6/2). Increasing number of grey clay streaks from 9.95 m	0.42	10.12
Silt, clayey, light olive-grey (5Y5/1). Finely laminated, with scattered, very fine-grained sand lenses. Gradational base	0.19	10.31
Silt, pale yellowish brown (10YR6/2); micaceous. Clay and very fine-grained sand lenses (lenticular lamination)	0.14	10.45
Clay, light olive (5Y5/1). Finely laminated and finely interbedded with very fine grained sandy silt	0.03	10.48
Silt, compact, brownish grey. Finely laminated and well developed vertical joints	0.03	10.51
Silt, pale yellowish brown (10YR6/2), hard; common very fine-grained sand lenses.		

	Thickness m	Depth m
Micaceous at 10.68 and 10.73 m. Iron pan well developed in silt at 10.72 . Gradational base, with increasing very fine-grained sand content	0.34	10.85
Silt, clayey brownish grey (5YR5/1). Occasional lenses of very fine-grained sand	0.13	10.98
Silt, brownish grey to pale olive-grey (5YR5/1 to 5Y5/1). Occasional lenses of very fine grained sand as above	0.67	11.65
Clay, silty, olive-grey (5Y5/1). Finely laminated	0.03	11.68
Silt, clayey, brownish grey (5YR5/1): micaceous	0.08	11.76
Clay, silty, hard, dark yellowish orange (10YR5/6) with slight iron pan developed	0.04	11.80
Silt, strongly ironstained in parts. Occasional lenses of very fine-grained sand up to 5 mm thick. Sharp base	0.52	12.32
Sand, very fine-grained, silty, dark yellow brown (10YR4/6) with dark brown iron pan at the top. Fining downwards through micaceous, very fine-grained, sandy silt to slightly very fine-grained sandy clay at 12.43 m. Sharp base	0.16	12.48
Silt, very fine-grained sandy, moderate yellowish brown (10YR5/4) with a dark brown iron pan from 12.60 to 12.62 m and at 12.66 m. Gradational base	0.49	12.97
Clay, silty, slightly very fine-grained sandy, light brownish grey (5YR6/1)	0.05	13.02
Silt, brown. Fining downwards into grey clay. Very thin lenses of very fine-grained sand throughout	0.04	13.06
Clay, olive-grey (5Y5/1). Grades into clayey silt, olive-grey; micaceous from 13.10 m	0.37	13.43
Clay, brownish grey	0.05	13.48
Sand, very fine-grained, silty with dark brown iron pan	0.02	13.50
Clay, grey (N6) with occasional very fine-grained sand streaks	0.02	13.52
Sand, very fine-grained with dark brown iron pan. Grades into very fine-grained sandy silt, yellowish brown (10YR5/2)	0.22	13.74
Silt, clayey, olive-grey. Finely laminated. Very fine-grained sand streaks throughout, buff and greenish grey (5Y6/1 and 5GY4/1)	0.13	13.87
Silt, clayey, very fine-grained sandy, brownish grey (5YR5/1). Laminated	0.15	14.02
Silt, clayey. Finely laminated. Very fine-grained sand laminae, buff (5Y6/1). Gradational base	0.12	14.14
Sand, very fine-grained, clayey, silty, yellowish brown, becoming dark greenish grey (10YR5/2 becoming 5GY3/1); micaceous. Cross-stratified, with interbedded very fine-grained sandy silt. Decrease in clay content down to 14.31 m below which clay is absent. Sharp erosive base	0.34	14.48
Silt, clayey, very fine-grained sandy, olive grey (5Y5/1); micaceous. Gradational base	0.05	14.53
Sand, very fine-grained, very silty; micaceous. Cross-stratified. Sharp erosive base channelled into bed beneath	0.07	14.60

	Thickness m	Depth m
Silt, clayey, compact, brownish grey (5YR5/1) with weak iron pan development. Grades down into silty very fine-grained sand, dark-olive grey (5Y3/1); micaceous, laminated	0.18	14.78
Silt, clayey, olive (5Y5/1 to 10YR5/2). Grades down into silty clay with scattered very fine grained sand and with occasional buff (5Y6/1) very fine grained sand streaks. Sharp base	0.35	15.13
Sand, very fine-grained, clayey and silty, dark yellowish brown and olive-grey (10YR3/2 and 5Y5/1); slightly micaceous. Finely interbedded with clayey, fine-grained, sandy silt. Reworked silty clay clasts at 15.42 to 14.43 m and 15.46 to 15.48 m	0.38	15.51
Sand, very fine-grained, silty; occasional carbonaceous remains. Cross-stratified, with buff and dark greenish grey, very fine-grained sand laminae. Well defined, erosive channel scour at base	0.07	15.58
Silt, clayey, olive-grey (5Y4/1). Finely laminated. Grades rapidly into fine-grained sandy silt, dark greenish grey (5GY4/1) with clay clasts up to 10 mm diameter. Abundant interlaminated very fine-grained sand, with a cross-stratified lens from 15.80 to 15.89 m. Gradational base	0.44	16.02
Sand, very fine-grained, silty, buff and light greenish brown; very micaceous. Common silty clay clasts. Gradational base, with increasing silt and clay content	0.12	16.14
Clay, silty, olive-grey (5Y4/1). Buff very fine-grained sand lenses increasingly abundant with depth and merging into a silty fine-grained sand at 16.21 m. Erosive irregular base	0.22	16.36
Clay, silty, brownish (5YR5/1). Abundant 1 to 2 mm lenses of buff, very fine-grained sand. Finely laminated. Grades down into very fine-grained sandy silt with irregular shaped clay clasts and buff silt laminae up to 10 mm thick	0.29	16.65
Silt, clayey, olive grey (5Y5/1). Grades into very fine-grained sandy, clayey silt with burrows 1 mm in diameter infilled with very fine-grained buff sand. Disseminated pyrite at 16.66 m	0.15	16.80
Clay, very silty, olive-grey. Many buff, very fine-grained sand partings up to 10 mm thick	0.29	17.09
Silt, very fine-grained sandy, olive-grey. Gradational base	0.03	17.12
Silt, clayey, brownish grey (5YR5/1). Grades into very silty clay with very fine-grained sand partings. Gradational base	0.49	17.61
Silt, very fine-grained sandy and clayey, olive grey (5Y5/1). Grades into buff (5Y6/1) silty, very fine-grained sand at base	0.08	17.69
Silt, clayey, brownish grey (5YR5/1). Occasional buff and very dark greenish brown, very fine-grained sand lenses	0.36	18.05
Core loss	2.63	20.68
Silty, clayey, olive grey (5Y5/1). Buff and dark greenish grey, very fine-grained sand lenses throughout. Becoming more sandy with depth. Gradational base	0.19	20.87

	Thickness m	Depth m
Sand, very fine-grained, silty brownish grey (5YR5/1). Grades into greenish grey (5GY5/1) clayey, very fine-grained sand containing *Venericardia* at 20.95, 21.01 and 21.13 m. Gradational base	0.38	21.25
Silt, clayey, olive-brown. Abundant buff, silty very fine-grained sand laminae. Pyritised twig 80 mm in diameter at 21.31 m. Gradational base	0.07	21.32
Clay, very silty, olive-grey (5Y5/1). Finely laminated. Abundant, very fine-grained sand lenses and silt streaks. Calcareous nodule 50 mm in diameter at 22.31 m	1.22	22.54
Clay, very silty, olive-grey (5Y5/1). Finely laminated as above. Slightly micaceous. Pyrite nodules at 22.55, 22.88 and 23.07 m	1.50	24.04
Septarian nodule with open calcite-lined fissures	0.14	24.18
Clay, silty, olive-grey (5Y5/1). Finely laminated. Gradually becoming increasingly silty with abundant silt streaks and occasional very fine-grained sand lenses. Gradational base	0.86	25.04
Silt, clayey, olive-grey (5Y5/1). Finely laminated	0.21	25.25
Silt, clayey as above, with abundant very fine-grained sand lenses and streaks. Grades into very fine-grained sandy clayey silt	0.42	25.67
Clay, very silty, as above. Becoming more sandy and silty towards the base (abundant microfauna including *Echinocythereis sp.*, '*Cytheridea*' *sp. nov.*, molluscs common, *Ringicula sp.*, *Adeorbis sp.*, Turridae, *Camptoceratops sp.*, between 22.00 and 26.00 m, unit 16)	0.61	26.28
Sand, very fine-grained, clayey, silty, dark greenish grey (5GY5/1) and buff (5Y6/1)	0.41	26.69
Sand, very fine grained, silty as above. Gastropods at 26.96 m (common microfauna including *Gaudryina hiltermanni* Meisl; molluscs common including *Corbula globosa*, '*Striarca*' *wrigleyi*, between 26.00 and 27.00 m, Unit 15)	0.30	26.99

LONDON CLAY

	Thickness m	Depth m
Clay, very silty, olive-grey (5Y5/1). Finely laminated. Grades into very fine-grained sandy silt with lenses of dark greenish grey, very fine-grained sand. Gradational base	0.19	27.18
Clay, very silty as above, with occasional very fine-grained sand partings up to 2 mm thick. Finely laminated buff and dark greenish grey (5Y6/1 and 5GY5/1), very fine-grained sand from 27.65 to 27.67 m; base burrowed. Gradational base	0.94	28.12
Clay, silty, olive-grey; slightly micaceous, finely laminated. Pyritised vertical burrow at 28.86 m	1.43	29.55
Clay, silty, olive-grey. Silt streaks and scattered dark greenish grey, very fine grained sand partings up to 20 mm thick. Less silty from 30.45 to 30.52 m. Increasing streaks and partings towards base [Core loss between 29.63 and 29.68 m]	1.52	31.07

	Thickness m	Depth m
Silt, clayey, as above; slightly micaceous. Abundant silt streaks. Decreasing silt content from 31.68 m. Gradational base	1.13	32.20
Clay, very silty as above. Dark greenish grey, very fine-grained sand band 5 mm thick at base	0.20	32.40
Clay, very silty as above. Lenses of dark greenish grey, very fine-grained sand up to 5 mm thick. Gradational base	0.91	33.31
Silt, very clayey as above. Buff silt streaks	0.10	33.41
Sand, very fine-grained, silty, dark-grey green (5GY4/1); micaceous. Irregular angular clay clasts up to 5 mm across. Erosional base	0.01	33.42
Silt, clayey, olive; micaceous. Finely laminated	0.04	33.46
Clay, very silty as above. Buff, very fine-grained sand partings up to 10 mm thick occur at about 0.1 m intervals. Gradational base	0.83	34.29
(common foraminifera and ostracods, *Echinocythereis sp.*, *Cytheridea sp.*, molluscs common including pteropods; *Spiratella tutelina*, *S. mercinensis*, between 27.00 and 34.00 m, Unit 14)		
Silt, clayey (silt approaches very fine-grained sand grade), olive (5Y5/1). Finely laminated. Mica content decreases with depth. Greenish grey lenses and streaks of very fine-grained sand	0.67	34.96
Sand, very fine-grained, silty, alternate buff and grey green (5Y6/1 and 5GY5/1) laminae	0.07	35.03
Silt, clayey, very fine-grained sandy. Very fine-grained sand partings up to 10 mm thick	0.03	35.06
Silt, clayey, olive (5Y5/1); micaceous. Finely laminated. Increasing silt content from 35.51 m. Scattered silty streaks throughout	0.63	35.69
Sand, very fine-grained, clayey, silty, olive grey; slightly micaceous. Finely laminated. Lenses of buff and greenish grey very fine sand from 5 to 30 mm thick at about 0.08 m intervals. Low sand content from 36.67 m and becoming less silty and slightly less micaceous. *Teredo* common from 36.45 to 36.57 m	1.84	37.53
Silt, clayey, olive, subordinate very fine-grained sand content increasing with depth; micaceous, finely laminated. Partings of buff and dark greenish grey, very fine-grained sand at 37.68 and 37.71 m	0.18	37.71
Silt, clayey as above; very micaceous from 37.71 to 37.83 m. Gradational base	0.52	38.23
Clay, very silty, olive. Few buff silt streaks. Slightly micaceous, well laminated. Becoming less silty towards the base	1.57	39.80
Clay, silty, silt streaks absent. Pyritised wood at 39.81 to 39.85 m. Gradational base (microfauna and macrofauna relatively sparse between 34.00 and 40.00 m, Unit 13)	0.59	40.39
Silt, clayey, olive-grey (5Y5/1). Gradational base	0.13	40.52
Clay, silty, very fine-grained sandy. Dark yellowish grey (5Y6/2). Buff and greenish grey, very fine-grained sand lenses. Gradational base	0.26	40.78

	Thickness m	Depth m
Clay, silty, very fine-grained sandy, olive (5Y5/1); slightly micaceous. Few very fine-grained sand lenses. Becoming increasingly silty	1.24	42.02
Sand, very fine-grained; micaceous and glauconitic. Silty clay bed from 42.02 to 42.06 m. Cross- stratified, with buff and grey green laminae inclined at 15° to 20°	0.05	42.07
Silt, very clayey, olive grey; micaceous. Occasional very fine-grained sand lenses. Gradational base	0.14	42.21
Clay, silty as above. Buff silt streaks and very fine grained sand partings	1.66	43.87
Silt, clayey, very fine-grained sandy, olive grey (5Y4/1); slightly micaceous. More clayey around 45.00 to 45.54 m then very silty with abundant buff silt and dark greenish grey, very fine-grained sand partings	1.98	45.85
Silt, very clayey, olive-grey as above. Occasional small scale fining-upwards rhythms about 0.02 m thick; gradations from a very fine sand to a silty clay	1.24	47.09
Clay, very silty (variable silt content), olive-grey; scattered mica flakes. Finely laminated. Small septarian nodule from 48.88 to 48.98 m. Increasing silt and mica content from 49.17 to 50.75 m. Gradational base	3.66	50.75
Clay, silty, olive-grey as above	0.03	50.78
Silt, clayey, light olive-grey (5Y5/1)	0.03	50.81
Clay, silty, olive-grey. Concentration of pyrite nodules of 10 mm maximum diameter at 50.86 m. Gradational base	0.09	50.90
Silt, clayey, olive-grey, slightly micaceous. Scattered buff and dark greenish grey, very fine-grained sand lenses and streaks. Gradational base, with increasing silt content	2.74	53.64
Silt, clayey, light brownish olive (5Y4/1) with glauconite pellets (of fine-grained sand grade) from 53.66 m. Common dark green and buff, very fine-grained sand laminae and buff silt streaks. Slightly less silty from 53.75 m	0.37	54.01
Clay, very silty, olive-grey; slightly micaceous, finely laminated. Buff silt streaks decreasing in abundance with depth	1.61	55.62
Clay, silty, olive-grey. Highly laminated. Numerous buff silt streaks	2.51	58.13
Clay, very silty as above. Burrows up to 7 mm diameter with infillings of buff, very fine-grained sand. Silt streaks increasing from 59.55 m. Streaks of glauconite and/or chlorite in very silty clay at 60.53 to 60.55 m	2.42	60.55
Clay, very silty as above. Few silt streaks. Bioturbated	2.74	63.29
Silt, clayey, olive-grey, with scattered glauconite pellets	0.20	63.49
Clay, very silty, olive-grey, slightly micaceous. Bioturbated, with relict primary lamination. Increasing silt with depth. [Core of poor quality from 65.43 to 65.60 m] (microfauna fairly abundant including *Cibicides* gr. *ungerianus* (d'Orbigny), *Anomalina sp.*, *Uvigerina batjesi* Kaaschieter, *Turrilina alsatica* Andreae, *Stilostomella spp.*,	2.13	65.62

	Thickness m	Depth m
Praeglobobulimina ovata (d'Orbigny), *Marginulina enbornensis* Bowen, *Cibicides westi* Howe, common at 49–65 m. *Pseudoclavulina anglica* Cushman and *Nodosaria minor* Hantken present at 60–65 m. *Terebratulina wardenensis* Elliot present at 57–59 m. Fauna between 40.00 to 65.00 m: Unit 12)		
Silt, clayey, olive-grey; micaceous. Scattered glauconite pellets. Dark greenish grey, very fine-grained sand lenses throughout	0.15	65.77
Clay, very silty, olive-grey; slightly micaceous and glauconitic as above	0.81	66.58
Silt, clayey, very fine-grained sandy, olive grey. Glauconitic pellets (very fine sand grade) common. Buff, very fine-grained sand lenses up to 5 mm thick occur throughout	0.32	66.90
Silt, clayey, slightly very fine-grained sandy, olive-grey; slightly micaceous. Buff and dark grey green, very fine sand lenses in decreasing number with depth	0.31	67.21
Silt, clayey, olive-grey; slightly micaceous with scattered lenses as above	0.41	67.62
Clay, silty, olive-grey tending to greyish brown. Finely laminated and partially bioturbated. Becoming more silty. Septarian nodule from 69.00 to 69.20 m. Colour changes back to olive-grey beneath the nodule	2.03	69.65
Silt, clayey, olive grey; occasional glauconite pellets from 70.10 to 70.40 m. Occasional dark greenish grey, very fine-grained sand lenses up to 5 mm thick. Becoming more clayey	1.49	71.14
Clay, very silty, olive-grey becoming yellowish brown. Increased compaction below 71.50 m. Slightly micaceous; finely laminated. Small septarian nodule at 72.22 m about 30 mm diameter	1.28	72.42
Core loss	1.66	74.08
Clay, very silty, light grey-brown (5YR4/1). Finely laminated. Occasional dark grey green, very fine grained sand streaks	0.36	74.44
Silt, very clayey, olive-grey (5Y4/1) with abundant buff silt streaks and greenish grey, very fine-grained sand lenses as above	2.04	76.48
Core loss	1.87	78.35
Silt, very clayey, brownish grey (5YR4/1); common mica flakes. Occasional greenish grey, very fine-grained sand streaks	0.79	79.14
Septarian nodule	0.10	79.24
Clay, very silty, olive-grey. Occasional glauconite pellets. Finely laminated and partially bioturbated: burrows 1 mm diameter infilled with grey, fine-grained sand and numerous 0.2 mm diameter burrows infilled with greenish blue silt. Some buff silt streaks and very fine-grained sand lenses (Microfauna less abundant. *Cibicides spp.*, *Spiraplectammina carinata* (d'Orbigny) s.1. commonest. *Ammodiscus sp.*, and *Glomospira sp.*, characteristic. Molluscs very scarce.	2.24	81.48

	Thickness m	Depth m
Between 65.00 and 81.00 m: Unit 11)		
Silt, very clayey, brownish grey (5YR4/1); very micaceous to 82.23 m. Finely laminated. Frequently dark green, pyritous very fine sand lenses becoming less common below 83.40 m. Burrows infilled with buff very fine-grained sand	4.12	85.60
Silt, very clayey, olive-grey. Very few silt lenses. Scattered pyrite nodules up to 10 mm diameter from 85.80 (to 87.73 m)	1.92	87.52
Clay, very silty, yellowish brown (10YR5/2). Well laminated. Occasional burrows up to 0.5 mm diameter infilled with greenish silt (Abundant microfauna including *Cibicides spp.*, *Spiroplectammina carinata* s.l., *Alabamina wilcoxensis* Tolmin between 81.00 and 88.00 m: Unit 10)	0.47	87.99
Clay, very silty, light brownish grey (5YR4/1). Well laminated; with burrows as above	0.92	88.91
Core loss	1.06	89.97
Silt, very clayey, olive-grey (5Y4/1). Very abrupt colour change at base with connate water in the core at the junction	2.70	92.67
Clay, very silty, yellowish brown (10YR5/2). Well compacted with brittle fracture. Abundant foraminifera. Well laminated and partially bioturbated. Small cementstone nodule 60 mm long at 95.01 m	4.90	97.57
Clay, silty as above, olive-grey (5Y4/1)	0.13	97.70
Clay, silty, (reduced) blue colouration (5B5/1). Pyritised carbonaceous material and *Teredo* fragments with associated 20 mm diameter burrows	0.02	97.72
Clay, silty as above. Scattered burrows infilled with blue-green silt. Scattered foraminifera. A single burrow 10 mm diameter infilled with buff silty, very fine-grained sand at 99.02 m	2.78	100.50
Septarian nodule	0.08	100.58
Clay, silty, brownish grey (5YR4/1), hard and compact with brittle fracture. Occasional burrows up to 0.5 mm infilled with bluish silt. Well laminated and partially bioturbated. Scattered foraminifera appear to be concentrated in more silty patches at 102.87 and 102.98 m. Gastropods at 103.47 m	2.98	103.56
Clay, silty as above, olive-grey (5Y4/1)	0.60	104.16
Silt, very clayey, olive-grey (5Y4/1); slightly micaceous. Excessively fissured giving a crumbly texture. Calcareous nodule 20 mm diameter at 104.30 m	0.14	104.30
Clay, very silty, olive-grey (5Y4/1); micaceous, crumbly as above. Scattered foraminifera and occasional burrows 0.2 mm across infilled with greenish silt	0.23	104.53
Clay, very silty, brownish grey to olive (5YR4/1 to 5Y4/1), stiff	0.09	104.62
Clay, very silty, brownish grey (5YR4/1). Crumbly texture. Scattered foraminifera	0.55	105.17

	Thickness m	Depth m
Clay, silty as above. Becoming slightly more silty at base	0.90	106.07
(Very abundant foraminifera including *Osangularia plummerae* Brotzen, *Nodosaria spp.*, *Lenticulina spp.*, *Marginulina enbornensis*, *Karreriella danica* (Franke), *Spiroplectammina carinata* s.l., *Cibicides westi*, *C. spp.*, *Pseudoclavulina anglica*, ostracods include *Trachyleberidea sp.*, *Cytherella sp.*, *Trachyleberis sp.* Planktonic foraminifera common. Molluscs very scarce. Between 93.00 and 106.00 m: Unit 9)		
Clay, very silty, olive-grey (5Y4/1); slightly micaceous. Pyritised burrows up to 1 mm diameter at 106.44 m	0.85	106.92
Clay, silty, olive-grey (5Y4/1). Scattered foraminifera	0.12	107.04
Clay, very silty, olive-brown. Occasional buff silt partings. Scattered foraminifera. Gradual colour change to olive-grey at 107.64 m and becoming slightly harder than above. Burrows infilled with greenish (?pyritous) silt up to 10 mm diameter and 60 mm long occur throughout. At 110.85 m, abundant 0.2 mm diameter burrows on a bedding surface with scattered, very fine-grained sand particles	4.24	111.28
Clay, very silty, brownish grey (5YR4/1). Hard and brittle. Scattered foraminifera and common burrows infilled with greenish silt. Bivalve at 112.50 m	3.18	114.46
Clay, very silty, less compacted than above. Few foraminifera and abundant burrows as above. Siltstone nodule (?phosphatic) at 114.36 m; with pyritised 0.5 mm diameter burrows	0.52	114.98
Clay, very silty, hard and brittle, brownish grey (5YR4/1). Rare foraminifera; burrows as above	0.45	115.43
Clay, very silty as above, with scattered buff silt streaks	0.03	115.46
Clay, very silty, olive-grey (5Y4/1). Bioturbated. Higher silt content from 116.75 to 117.40 m	3.93	119.39
Clay, very silty, excessively fissured giving a crumbly texture, olive-brown (5YR4/1). Light brown siltstone nodule at 119.83 m; darker core suggests phosphate content	0.45	119.84
Clay, silty, brownish grey (5YR4/1), soft. Small pyrite nodules 10 mm in diameter at 120.02 m	0.44	120.28
Clay, very silty, olive-brown (5YR4/1). Crumbly texture. Light brown (phosphatic) siltstone nodule at 120.37 m (40 × 30 mm)	0.40	120.68
Core loss	1.76	122.44
Silt, very clayey, olive-grey, crumbly texture. *Balanocrinus* stem 30 mm long at 122.74 m. Well defined base	0.55	122.99
(A similar fauna to above but very rare *Osangularia plummerae*. *Balanocrinus subbasaltiformis* present at base. Between 106.00 and 122.00 m: Unit 8)		

	Thickness m	Depth m
Clay, very silty, olive-brown (5YR4/1), hard and brittle. Crinoid fragments at 123.01 m. Bioturbated, with irregular patches of very fine grained sand. Common small (2 to 10 mm diameter) burrows infilled with greenish silt	1.54	124.53
Silt, very clayey, olive-grey. Occasional bioturbated buff silt patches. Burrows as above, becoming more abundant	2.06	126.59
(Reduced microfauna and pyritised diatoms common, between 123.00 and 125.00 m: Unit 7)		
Core loss	0.42	127.01
Clay, silty, olive-grey (5Y4/1), crumbly texture. Frequent burrows up to 1 mm diameter. Gradational base	0.18	127.19
Silt, very clayey, olive-grey, crumbly as above. Burrows up to 30 mm diameter	0.23	127.42
Clay, very silty, olive-brown (5Y4/1), hard and brittle. Buff siltstone (phosphatic) nodule at 127.56 m	0.28	127.70
Clay, very silty, olive-brown (5YR4/1)	0.49	128.19
Silt, clayey, olive-grey, hard and brittle; with abundant small burrows as above. Prominent vertical joint	0.99	129.18
Clay, very silty, excessively fissured giving a crumbly texture. Burrows as above	0.17	129.35
Core loss	0.15	129.50
Clay, very silty, olive-grey. Brittle fracture with a prominent joint running from the top down to 129.93 m. Becoming more silty downwards	1.26	130.76
(Common foraminifera; microfauna becomes scarce towards base, between 125.00 and 130.00: Units 5/6)		
Silt, very fine grained sandy, very clayey, olive-grey. Scattered burrows filled with greenish, very fine-grained sand. Well jointed	0.44	131.20
Clay, very silty, olive-grey. Scattered burrows as above. Well jointed	0.42	131.62
Clay, very fine-grained sandy, silty, olive-grey. Bioturbated, with irregular concentrations of very fine-grained sand. Pyritised wood with abundant boring tubes filled with pyritous silt	0.47	132.09
Clay, very fine-grained, sandy, silty. Bioturbated, with irregular patches of buff, very fine-grained sand. Scattered burrows filled with greenish silt throughout. Very fine sand content increases with depth	0.21	132.30
Sand, very fine, clayey, olive-grey; slightly micaceous. Gradational base	1.51	133.81
Silt, clayey, olive-grey. Gradational base	0.39	134.20
Clay, very fine-grained sandy, silty, olive-grey; micaceous. Bioturbated with partially disturbed, dark greenish grey, very fine-grained sand laminae up to 5 mm thick. Increasing number of buff, very fine-grained sand streaks from 134.82 m	1.04	135.24
Silt, clayey, olive-grey. Scattered buff silt streaks and greenish grey, very fine-grained sand streaks. Gradational base	1.08	136.32

	Thickness m	Depth m
Clay, very silty, olive-brown. Bioturbated, with irregular silt patches and streaks	0.68	137.00
Core loss	3.15	140.15
Clay, silty, olive-brown. Increasing silt content towards the base	0.31	140.46
Clay, very silty with dispersed very fine-grained sand	0.13	140.59
Clay, very silty, grey-brown. Hard with brittle fracture	0.15	140.74
Clay, silty, olive-grey. Abundant silt streaks throughout. Silty, very fine-grained sand lenses up to 5 mm thick. Occasional concentrations of white silt infilled burrows up to 1 mm diameter on bedding surfaces. Scattered pyrite nodules up to 20 mm diameter	1.09	141.83
(Abundant arenaceous foraminifera: *Ammodiscus sp.*, *Rheophax sp.*, in upper part; *Spiroplectammina carinata* s.l. and calcareous foraminifera, *Lenticulina spp.* and planktonic forms in lower part. Between 130.00 and 141.00 m: Units 3/4)		
Silt, clayey, very fine-grained sandy, olive grey. Few silt streaks and very fine-grained sand lenses. Burrows up to 1 mm diameter filled with greenish silt	1.53	143.36
Silt, clayey, olive grey; micaceous (silt is coarse approaching very fine sand grade). Pyritised root at 143.57 m. Slightly glauconitic or chloritic clay from 144.02 to 144.10 m and at 145.37 m. Buff silt streaks from 144.08 m downwards	2.08	145.44
Silt, very fine-grained sandy, clayey, olive-grey; micaceous. Abundant silt and very fine-grained sand streaks. Scattered chitinous fragments (?fish scales)	0.55	145.99
Silt, clayey, olive-grey; micaceous. Bioturbated, with common buff silt streaks and occasional very fine-grained sand streaks	0.81	146.80
Silt, very fine-grained sandy, clayey, olive-grey (5Y4/1); micaceous. Bioturbated. Abundant disseminated pyrite on some bedding surfaces	0.31	147.11
Silt, clayey, very fine-grained sandy (sand and silt about equal in amount); micaceous. Bioturbated in part with abundant burrowed, buff and grey, very fine-grained sand lenses and silt streaks. Partially pyritised wood at 148.04 m. Becoming less sandy at the base	1.15	148.26
Silt, clayey, olive-grey; micaceous. Bioturbated with some very fine-grained sand and buff silt streaks. Becoming more clayey at the base	0.16	148.42
Silt, very clayey as above. Concentrations of small burrows (0.5 mm diameter) infilled with very pale silt and very fine-grained sand at 148.87 m. Black rounded flint pebble (30 × 20 mm) at 148.76 m	0.45	148.87
Silt, clayey, very fine-grained sandy, olive-grey: micaceous. Bioturbated with relict silt and very fine-grained sand laminae. Abundant very fine-grained sand lenses and patches	0.22	149.09

	Thickness m	Depth m
Sand, very fine-grained, silty, olive-grey. Burrows infilled with buff, very fine-grained sand. Pyrite nodule (10 × 10 mm) at 149.16 m	0.18	149.27
Silt, clayey and very fine-grained sandy, olive and bluish grey. Silt streaks and burrows. Decreasing fine sand content downwards. Burrows up to 30 mm diameter infilled with very fine-grained sand	1.12	150.39
Sand, very fine-grained, silty, greenish olive (5Y4/1). Patchy distribution of bioturbated very fine-grained sand. Irregular concentration of 0.5 mm burrows	0.29	150.68
Silt, very fine grained sandy. Abundant very fine grained sand streaks, becoming more clayey with depth and grading into clayey silt with occasional very fine grained sand streaks. Increasing very fine-grained sand content from 151.22 m	0.71	151.39
Sand, very fine-grained, silty, greenish grey. Bioturbated	0.17	151.56
Silt, clayey, very fine-grained sandy, greyish green (5GY4/1). Bioturbated, with burrows up to 5 mm diameter and infilled with very fine-grained sand and small (0.5 mm) burrows infilled with greyish silt	0.60	152.16
Clay, silty, grey to greenish blue (5B5/1). Abundant 1 mm diameter burrows filled with greenish silt. Gradational base, with increasing very fine-grained sand content	0.67	152.83
Sand, very fine-grained, silty, moderate greyish green (5GY5/1). Bioturbated, with white calcareous burrows concentrated in irregular patches and larger burrows (5 mm) infilled with pale olive silt	0.43	153.26
Silt, clayey, very fine-grained sandy, grey green. Bioturbated with burrows as above. Common silt streaks and irregular coarse silt patches. Becoming more clayey from about 153.99 m	1.02	154.28
Sand, very fine-grained silty, mottled olive and greyish green (5Y4/1 and 5GY5/1); micaceous	0.26	154.54
Silt, clayey, very fine-grained sandy, greyish green changing to olive-grey to 154.70 m. Becoming more sandy from 154.81 m	0.38	154.92
Sand, very fine-grained, olive-grey (5Y4/1) with dark greenish grey more sandy patches. Bioturbated, with burrows as above	0.84	155.76
(Very sparse arenaceous foraminifera and very abundant pyritised diatoms; *Coscinodiscus sp.*, *Triceratium sp.*, between 141.00 and 155.00 m, Unit 2)		
Sand, very fine-grained, dark greenish olive (5GY4/1). Finely laminated (1 to 2 per mm)	0.02	155.78
Sand, very fine-grained, silty as above. Homogeneous	0.12	155.90
Sand, very fine-grained, dark olive (5Y3/1). Finely laminated (1 to 2 per mm). Well defined base	0.28	156.18
Sand, very fine-grained, clayey, olive-grey (5Y4/1). Bioturbated, becoming less clayey towards the base	0.17	156.35

	Thickness m	Depth m
Sand, very fine-grained; scattered glauconite grains from 156.38 to 156.96 m. More clayey towards the base	0.61	156.96
Sand, very fine-grained, clayey, olive-grey (5Y4/1). Bioturbated	0.07	157.03
Sand, very fine-grained and fine-grained, slightly clayey; olive-grey. Bioturbated	0.25	157.28
Sand, very fine-grained, clayey, bluish green (5B4/1). Strongly bioturbated with numerous burrows infilled with buff, very fine-grained sand	0.04	157.32
Sand, very fine-grained and fine-grained, silty; dark grey. Strongly bioturbated	0.77	158.09
Silt, very fine-grained sandy and clayey. Scattered broken shells from 158.19 m. Pyritised burrows at 158.20 m	0.16	158.25
Siltstone, very fine-grained sandy, grey to pale olive. Slightly calcareous, carbonaceous fragments common	0.09	158.34
Silt, very fine-grained sandy, clayey. Scattered glauconite pellets from 158.34 m. Bioturbated, with nests of bivalves (Corbula). Increasingly very fine-grained sandy from 158.96 m. Partially pyritised wood fragments	0.64	158.98
Sand, fine-grained, silty, grey-green (5Y4/1 to 5GY4/1). Common glauconite pellets. Irregular silty clay clasts and occasional subrounded and subangular black flint pebbles up to 10 mm diameter	0.06	159.04
(Sparse microfauna at the top, but much shell debris in lower parts. Astarte present. Microfauna includes Globulina spp, Protelphidium sp.3 Murray & Wright, between 155.00 and 159.00 m: Unit 1)		
Septarian nodule, (bleached) pale olive, with pockets of carbonaceous material. Faint relict lamination	0.06	159.10
Clay, silty. Bioturbated, with occasional burrows up to 10 mm diameter infilled with grey green, very fine-grained glauconitic sand. 10 mm bed containing abundant small carbonaceous fragments at 159.14 m. Well defined base	0.09	159.19

?OLDHAVEN BEDS

	Thickness m	Depth m
Sandy, fine- to medium-grained, shelly, yellowish grey (5Y7/3). Abundant shelly material; bivalves with concave side upwards. Black rounded flint pebbles (30 × 20 mm) at 159.22 m. Abundant dark green glauconite pellets	0.06	159.25
Core loss	1.60	160.85

WOOLWICH BEDS

	Thickness m	Depth m
Sand, medium-grained, light grey (N7). Scattered black (N3) medium-grained sand laminae up to 10 mm thick. Sand grains subangular to subrounded	0.28	161.13
Sand, medium-grained as above, dark grey (N5) with ?carbonaceous streaks from 161.13 to 161.19 m and from 161.29 to 161.35 m. Core broken and crumbly	0.22	161.35

	Thickness m	Depth m
Sand, medium grained, pale olive-grey (5Y6/1) with medium grey (N5) clay laminae up to 5 mm thick. Scattered broken shells at 161.35 m. Scattered carbonaceous material throughout. Occasional coarse sand grains	0.38	161.73
Sand, medium-grained, light olive-grey (5Y6/1), with abundant bivalves (Corbicula cuneiformis (J. Sowerby) between 160.85 and 162.00 m)	0.30	162.03
Sandstone, medium-grained. Homogeneous	0.18	162.21
Sand, medium-grained, dark brownish grey (10YR2/2) and banded light and dark grey (N7 and N3). Slightly clayey in part, normally in the darker grey bands. Occasional ?glauconite pellets	0.05	162.26
Core loss	2.86	165.12
Sand, medium-grained, greyish olive (10Y5/2). Sand grains are mainly subangular; occasional coarse-grained sand. Frequent (medium-grained sand grade) pyrite concretions. Scattered carbonaceous remains in the upper 0.15 m	0.61	165.73
Sand, medium-grained, pale grey. Occasional beds of carbonaceous, clayey, medium- and fine-grained sand with black, carbonaceous, very fine-grained, sandy clay wisps up to 3 mm thick	0.09	165.82
Core loss	2.17	167.99
Sand, medium-grained, yellowish grey and green. Frequent glauconite pellets	0.30	168.29
Sand, medium-grained, as above. Dark grey (N4) clay bands up to 20 mm thick. A reworked, partially rounded clay clast 40 mm in diameter at 168.33 m	0.10	168.39
Sand, medium-grained as above. Cross-stratified channel-fill, 50 mm deep with interbedded grey clay seams up to 20 mm thick; clay lining to base of channel	0.26	168.65
Sand, medium-grained, brownish grey. Scattered grey clay clasts throughout	0.11	168.76
Sand, medium-grained, greenish and yellowish grey. Clay laminae abundant (3 to 4 per 10 mm) from 168.76 to 168.80 m	0.20	168.96
Sand, medium-grained as above. Abundant glauconite pellets and rounded black flint pebbles; maximum pebble diameter of 20 mm	0.03	168.99
Sand, medium- and fine-grained, greenish grey. Becoming finer grained with depth	0.10	169.09
Sand, medium- and fine-grained, as above, black ?carbonaceous fine-grained sand laminae up to 10 mm thick throughout	0.06	169.15
Core loss	2.14	171.29

THANET BEDS

	Thickness m	Depth m
Sand, fine- and medium-grained, mottled dark greenish grey and pale grey (5GY4/1 to N3). Bioturbated; occasional fragile thin shelled molluscs. Scattered carbonaceous clayey streaks in the upper 0.45 m. Becoming finer grained and more silty with depth	2.23	173.52

	Thickness m	Depth m
Sand, fine-grained, silty, slightly clayey; greenish grey. Black very fine-grained sandy clay partings up to 30 mm thick at 174.12 and 174.20 m. Abundant bivalves at 173.90 m and at 174.00 to 174.12 m. Carbonaceous material at 174.30 m	0.95	174.47
Sand, very fine-grained, silty, pale grey. Bioturbated. Occasional black clay partings up to 10 mm thick	0.23	174.70
Sand, very fine-grained, silty, pale grey with yellowish grey patches. Bioturbated, with burrows up to 120 mm long and 15 mm across from 175.10 to 175.30 m. Scattered carbonaceous fragments throughout.	seen to 0.84	175.54
(Microfauna very sparse with *Ceratobulimina tuberculata* Brotzen. Mollusc fragments common including *Lucina sp.*, *Nucula sp.*, *Nuculana sp.*, between 171.29 and 175.00 m)		
Core loss	0.19	175.73
Borehole terminated		

SHEET TQ 88 NW

87 BGS Plumtree Hill, Castle Farm, Hadleigh Borehole [8083 8627]
Surface level + 50.20 m OD;
Date 1974

	Thickness m	Depth m
TOP SOIL	0.30	0.30
CLAYGATE BEDS		
Sand, fine-grained, silty, dark yellowish orange (10YR6/6)	0.10	0.40
Clay, silty, light brown (5YR6/4)	0.70	1.10
Clay, silty, medium yellow-brown (10YR5/5); laminae of fine sand with abundant mica	0.70	1.80
Clay, very silty, with septarian nodules; fissured	0.40	2.20
Clay, stiff with a few fine sand laminae, micaceous; selenite crystals	0.30	2.50
Clay, silty, sandy. Fine sand content decreasing to base. Selenite common	0.40	2.90
Clay, silty, sandy, medium brown (5YR5/5) with fine sand laminae, micaceous selenitic	0.70	3.60
Clay, sandy with scattered selenite crystals	0.20	3.80
Sand, fine-grained, clayey, micaceous, glauconitic, medium yellow-brown (10YR5/5), subangular flint	1.30	5.10
LONDON CLAY		
Clay, very silty, medium, yellow-brown (10YR5/3). Lenses and streaks of fine grained, micaceous sand. Scattered selenite crystals	2.20	7.30

	Thickness m	Depth m
Clay, very silty, light olive-grey (5Y5/1), with fine-grained sand streaks and laminae, micaceous	2.70	10.00
Silt, clayey, light olive-grey (5Y5/1) with beds of grey green micaceous, fine-grained sand	1.20	11.20
Clay, very silty with sand laminae and streaks	1.20	12.40
(Ostracods common including *Cytheridea primitia*, *Echinocythereis sp.* and *Paijenborchella sp.* between 7.90 and 11.50 m)		
Clay, very sandy, fine-grained, with beds and laminae of very fine-grained sand, dark green (5GY3/1). Becoming more sandy at base with beds of dark green and light grey very fine-grained sand in olive (5Y5/1) sandy clay, micaceous, finely laminated	3.00	15.40
Clay, very silty, with abundant very fine-grained sand partings. Bioturbated	1.10	16.50
Sand, clayey, very fine-grained, laminated, bioturbated. Grading downwards to very silty clay with micaceous sand lenses	0.80	17.30
Clay, very silty with very fine-grained sand lenses, laminae and streaks	0.50	17.80
Silt, clayey, light olive-grey (5Y5/1), micaceous	0.60	18.40
Clay, very silty, stiff, light olive-grey (5Y5/1)	seen to 1.60	20.00

TQ 88 NW

89 BGS Castle Farm, Hadleigh Borehole [8086 8646]
Surface level + 57.20 m OD;
Date 1973

	Thickness m	Depth m
HEAD		
Clay, sandy, silty with scattered flint pebbles	0.80	0.80
CLAYGATE BEDS		
Clay, sandy (fine), medium yellow-brown (10YR5/6)	0.40	1.20
Sand, fine-grained, clayey, greyish orange (10YR7/4) mottled greenish grey (5GY6/1)	0.30	1.50
Clay, silty, sandy (fine), moderate yellow-brown (10YR5/4) mottled light bluish grey (5B7/1)	0.30	1.80
Sand, clayey, fine-grained, finely laminated, micaceous sands and clayey silts, moderate yellow-brown (10YR5/4). Some thin laminae hardened by iron pan cementation. Many sandy units contain carbonaceous material	1.00	2.80

	Thickness m	Depth m
Clay, silty, stiff, medium yellow-brown (10YR5/2) with abundant fine sand streaks and lenses, micaceous fine sand beds. Fissuring	0.20	3.00
Silt, fine-grained sand, medium yellow-brown (10YR4/4) micaceous	0.90	3.90
Sand, fine-grained, silty, micaceous, medium yellow-brown (10YR5/2) with interbedded stiff, very silty clay containing abundant sandy streaks	0.20	4.10
Clay, stiff, silty, medium yellow-brown (10YR5/2) with lenses and laminae of very fine-grained sand, micaceous. Fissured	1.00	5.10
Septarian nodule	0.10	5.20
Clay, very silty, stiff, light olive-grey (5Y5/1) fissured. Finely laminated and scattered streaks of greenish grey (5GY5/1), micaceous, very fine-grained sand	1.00	6.20
Clay, very silty, stiff, light olive-grey (5Y5/1) fissured	0.10	6.30
Clay, very silty, stiff, micaceous with finely laminated pyritous and micaceous very fine-grained sand, and streaks	1.70	8.00
Silt, clayey, stiff, light olive-grey (5Y5/1), bedded. Abundant micaceous, very fine-grained sand laminae. Transitional base	0.30	8.30
Sand, very fine-grained, clayey with micaceous very fine sand streaks. Finely laminated	0.40	8.70
(*Turrilina brevispira* Dam, *Astarte sp.*, '*Cytheridea*' sp. nov. between 7.40 and 8.50 m)		
Silt, clayey with beds of laminated fine-grained sand	0.60	9.30
Sand, fine-grained, silty, greenish grey (5GY5/1) grading to light olive-grey (5Y5/1). Micaceous	0.60	9.90

LONDON CLAY

	Thickness m	Depth m
Silt, very clayey, light olive-grey (5Y5/1) with abundant micaceous, very fine-grained sand streaks and laminae	1.10	11.00
Clay, silty, finely laminated with sandy streaks; fissured, broken shells	3.40	14.40
Clay, very silty, stiff, with fine sand laminae and streaks	3.10	17.50
Silt, sandy, very fine-grained, micaceous, clayey, stiff with abundant, very fine sand streaks and laminae	0.80	18.30
Silt, very clayey with sand streaks and laminae	seen to 0.70	19.00

SHEET TQ 89 SW

37 BGS Hockley Heights (Gattens) Borehole [8176 9205]
Surface level + 74.09 m OD;
Date 1973

	Thickness m	Depth m
HEAD		
Sand, very fine-grained, silty with scattered flint pebbles, mottled orange-brown and pale grey (dug in trial pit)	1.50	1.50
Core loss	5.56	7.15
BAGSHOT BEDS		
Sand, very fine-grained, silty, brown (5YR5/6); slightly micaceous, with a thin iron pan development at 7.26 m	0.14	7.29
Sand, very fine grained, silty, moderate yellowish brown and dark yellowish orange (10YR5/4 and 10YR5/6); thin very fine-grained sand bands up to 2 mm thick, light brownish grey (5YR6/1). Gradational base	0.10	7.39
Sand, very fine-grained, silty, brown; interbedded with thin bands of clay, light grey, finely laminated, with sharp tops and bases	0.10	7.49
CLAYGATE BEDS		
Sand, very fine-grained, very silty, moderate yellowish brown (10YR5/4), slightly micaceous. Well developed iron pan from 7.49 to 7.52 m	0.11	7.60
Sand, very fine-grained as above [Only 0.39 m recovered]	1.56	9.16
Sand, very fine-grained, very silty as above. Fine lamination accentuated by differential iron staining. Slightly micaceous	0.31	9.47
Sand, very fine-grained, silty, yellowish brown and dark yellowish orange (10YR5/4 and 10YR6/6); laminated. Bed of clay, 20 mm thick, pale yellow brown (10YR6/2) containing carbonaceous flecks; the bed has a sharp top and base	0.10	9.57
Silt, slightly clayey with some very fine-grained sand, banded dark yellow-brown and dark yellowish orange (10YR5/4 and 10YR6/6). Light greenish grey (5GY6/1) clayey silt from 9.64 to 9.69 m	0.12	9.69
Silt, clayey, very fine-grained, sandy, light yellowish brown (19YR6/2). Beds of clay, light grey (N5) containing finely comminuted carbonaceous remains from 9.73 to 9.74 m and from 9.81 to 9.83 m; these beds have irregular but well defined bases and sharp planar tops. Small lenses of silty fine-grained sand occur within the clay beds	0.16	9.85
Silt, clayey, very fine-grained, sandy, yellowish brown as above; laminated. Bed of finely laminated silty clay, light greenish grey, from 9.91 to 9.94 m. Gradational base	0.10	9.95
Silt, clayey, very fine-grained, sandy, light yellowish brown (10YR6/2); micaceous, laminated. Clay seams, light grey, at 10.28 and 10.29 m	0.36	10.31
Silt, very fine-grained, sandy, light brown (5YR5/6); slightly micaceous	0.69	11.00

	Thickness m	Depth m
Sand, very fine-grained, silty with finely interbedded very fine-grained sandy silt and clayey silt, grey brown (5YR5/1) and yellow-brown (10YR5/4). Small scale cross-stratification in the more sandy beds, many of which are only 2 to 3 mm thick	0.38	11.38
Core loss	0.88	12.26
Sand, very fine-grained, very silty, brown [Poor quality core]	0.12	12.38
Sand, very fine-grained, silty, dark yellowish orange (10YR5/6) laminated	0.04	12.42
Clay, very fine-grained sandy, dark yellowish orange to brownish grey (10YR5/6 to 5YR5/1)	0.06	12.48
Clay, very fine-grained sandy, silty, soft, olive-grey (5Y4/1); slightly micaceous. Occasional firm beds of silty clay, olive-grey, up to 15 mm thick. Increasing silt content towards the base	0.39	12.87
Silt, very fine-grained sandy, clayey, olive-grey (5Y4/1); micaceous, finely laminated. Much of the 'sand' content may be silt grade	0.05	12.92
Silt, as above, with very fine-grained sand partings a few grains in thickness, buff (5Y6/1)	0.06	12.98
Silt, as above. [Core wet and broken]	0.09	13.07
Sand, very fine-grained clayey interbedded with silty clay, olive-grey (5Y4/1). Very fine-grained sand lenses from 13.17 m downwards. Large fine-grained sand lenses from 13.17 m downwards. Large round pyrite nodule (50 × 30 mm) at 13.15 m. Small pyrite nodule at 13.19 m. Gradational base (*Lingula sp.* between 12.68 and 13.40 m becoming abundant near base. Units 19 and 20)	0.58	13.65
Clay, very silty, olive (5Y4/1); micaceous. The clay approaches silt grade. Silty, very fine-grained sand streaks throughout, light olive (5Y5/1); incipient pyrite nodule at 14.70 m. Slickensided bedding planes at 14.50 m. Small scale coarsening-upwards cycles 50 mm thick grade from silty clay to clayey silt with reworked wispy clay pellets in the upper 20 mm	1.16	14.81
Mudstone, calcareous	0.07	14.88
Silt, clayey, olive-grey (5Y4/1); micaceous. Contains irregular patches of silt, greyish green (5GY4/1 to N3), as a result of bioturbation. Bed of hard silt, light to olive (5Y3/1) from 14.93 to 14.95 m has an irregular bioturbated top and a sharp base	0.07	14.95
Silt, very clayey, olive-grey (5Y4/1); slightly micaceous. Scattered very thin buff silt streaks. Becoming gradually more silty towards the base	0.25	15.20
Silt, clayey, olive, as above; micaceous, laminated. Becoming increasingly micaceous and crumbly at 15.48 . [Core broken from 15.48 to 15.54 m] (Pyrite moulds of molluscs, with *Corbula globosa* dominant; also *Camptoceratops prisca*; between 13.69 and 15.38: Unit 18 part)	0.34	15.54

	Thickness m	Depth m
Clay, very silty, olive (5Y4/1); slightly micaceous. Frequent buff (5Y5/1) silt streaks 4 mm thick. The silt lenses are often finely cross-stratified and comprise alternating buff and dark olive laminae (5Y5/1 and 5Y4/1). The silt lenses show varying degrees of reworking frequently containing wisps of dark olive clay and having a burrowed base. Microfaulting and loadcasting phenonomena at 17.05 m. Abundant silt streaks from 16.97 to 17.22 m and 17.29 to 17.33 m. Pyrite nodules at 16.63 and 17.07 m (Pyrite moulds of molluscs scarce to common, including *Nuculana sp.* and *Corbula globosa*, between 15.48 and 17.33 m, Unit 18 part)	1.79	17.33
Core loss	2.41	19.74
Sand, very fine-grained, silty, dark olive (5Y3/1). Sharp erosive base. [Core wet and broken from 19.74 to 19.80 m]	0.16	19.90
Silt, clayey, olive (5Y4/1); slightly micaceous. Lenses of very fine-grained sand, (5Y3/1), increasing towards the base	0.44	20.34
Sand, very fine-grained, very silty; slightly micaceous, finely laminated. Sharp erosive base. [Core wet and broken from 20.50 to 22.00 m]	0.40	20.74
Clay, very silty, olive-grey; fissured	0.06	20.80
Sand, very fine-grained, slightly silty, greenish dark olive; micaceous; common dark green glauconite pellets, laminated (about 2 per mm). Increasing silt content from 21.60 m to the base	1.22	22.02
Silt, very fine-grained sandy; finely laminated (1 to 2 per mm). Silty clay wisps increasing from 22.13 m and becoming discontinuous laminae up to 1 mm thick (reworked coarsening-upwards cycle). Scattered carbonaceous remains throughout. Gradational base	0.13	22.15
Clay silty, olive-grey. Abundant lenses of very fine sand. Well defined base	0.11	22.26
Clay, very fine-grained sandy, coarsening downwards into finely cross-stratified, very fine-grained sand with 1 to 2 laminations per mm	0.16	22.42
Clay, silty, olive-grey (5Y4/1). Abundant light and dark olive (5Y5/1 to 5Y3/1), very fine-grained sand and silt partings. Gradational base	0.15	22.57
Silt, clayey, dark to medium olive-grey (5Y4/1 to 5Y5/1); micaceous, finely laminated. Scattered silt streaks. Concentrations of mica flakes on some bedding plane surfaces. Grades gradually into very silty clay with cross-stratified silt and very fine-grained sand lenses, up to 20 mm thick. The lenses have marked erosive bases, occasionally burrowed, with bioturbated or sharp tops. Burrows up to 5 mm long filled with buff (5Y6/1) very fine-grained sand (*Lingula sp.*, common at 19.74 to 20.82 m; sporadic mollusc moulds (decalcified). Between 19.74 and 23.25 m: Unit 17)	0.68	23.25
Core loss	3.27	26.52
Clay, very silty, olive (5Y4/1); finely laminated	0.17	26.69

	Thickness m	Depth m
Clay, silty, with interbedded clayey very fine-grained sand from 26.69 to 26.70 m and from 26.83 to 26.86 m, and cross-stratified, very fine-grained sand lenses and thin beds up to 20 mm thick, greenish olive and buff	0.28	26.97
Silt, clayey, with scattered very fine sand grains, light olive (5Y5/1); abundant disseminated carbonaceous fragments	0.07	27.04
Silt, clayey, light olive (5Y5/1); slightly micaceous; abundant carbonaceous remains from 27.70 to 27.77 m. Common dark greenish grey, cross-stratified very fine-grained sand lenses and buff silt streaks throughout. Abundant small (1 mm diameter) burrows from 27.92 to 27.94 m. Small bivalves at 27.30 m. Gastropod (*Tibia*) at 27.25 m. Large (50 mm diameter) pyrite nodule at 27.43 m. Gradational base (A calcareous fauna including common molluscs *Ringicula sp.*, *Nuculana sp.*, Turridae and foraminifera, *Lenticulina spp.* dominant, between 26.52 and 27.80 m: Unit 16)	0.93	27.97
Silt, very fine-grained sandy, olive-grey (5Y4/1). Lenses of very fine-grained sand, buff (5Y6/1), finely cross-stratified. Becoming more clayey between 28.10 and 28.14 m	0.25	28.22
Silt, very fine-grained sandy, greenish olive. Homogeneous texture (Common molluscs including Turridae, *Ringicula sp.*, *Pachysyrnola sp.* and *Euspira sp.* Microfauna relatively restricted. Between 28.00 and 28.54 m: Unit 15 part)	0.32	28.54
Core loss	2.25	30.79

LONDON CLAY

	Thickness m	Depth m
Silt, very clayey, soft to firm, olive-grey (5Y4/1); micaceous. Dark green and buff silt and very fine-grained sand streaks	1.05	31.84
Silt, very clayey, stiff, olive-grey (5Y4/1). Homogeneous texture in the top metre, becoming more silty with depth and containing greenish silt laminae from 32.80 m	2.19	34.03
Silt, clayey, olive. Frequent lenses and streaks of buff and greenish silt and very fine-grained sand. Becoming more clayey from 34.50 m with an increasing number of lenses up to 20 mm thick. Sharp base	0.63	34.66
Silt, (approaching very fine-grained sand grade), clayey, olive-grey. Many greenish grey silt and very fine-grained sand streaks as above. Abundant glauconite pellets from 34.66 to 34.77 m. More clayey from 34.66 to 34.80 m. Gradational base	0.93	35.59
Silt, clayey, olive. Buff and dark grey-green (5GY2/1) very fine-grained sand and silt streaks. Streaks becoming less frequent from 36.50 m. Higher silt content from 36.03 to 36.40 m then decreasing to the base. Scattered shell fragments around 36.65 to 36.75 m. Slickensides along bedding surfaces from 37.77 to 38.00 m. Scattered pyrite nodules and pyritised wood at 38.25 m. Gradational base	2.84	38.43

	Thickness m	Depth m
Clay, very silty, olive-grey. Scattered buff silt streaks. Slickensides along bedding surfaces from 38.64 to 38.87 m. Increasing silt towards base	0.54	38.97
Silt, very clayey, olive-grey. Silt streaks increasing with depth and becoming thicker. Finely laminated silt beds up to 5mm thick towards the base. Well developed vertical jointing, some with groove markings. Slickensides along bedding surfaces from 39.00 to 39.14 m	0.56	39.53
Silt, clayey, olive (5Y4/1) (silt approaches very fine-grained sand grade). Abundant buff silt streaks and lenses. Bioturbated	0.49	40.02
Silt, very clayey, olive with silt laminae and finely laminated very fine-grained sand lenses up to 20 mm thick	seen to 0.19	40.21

SHEET TQ 89 SE

41 BGS Ashingdon Borehole [8557 9337]
Surface level +55.1 m OD;
Date 1972

	Thickness m	Depth m
TOP SOIL	0.30	0.30

TERRACE DEPOSITS

	Thickness m	Depth m
Sand, clayey, gravelly, dark yellowish orange (10YR6/6). Angular flints and rare quartzite fragments	0.90	1.20
Sand, medium coarse-grained, clayey, medium yellow-brown (10YR4/6) mottled dusky yellow (5Y6/4). Weathered	0.10	1.30

CLAYGATE BEDS

	Thickness m	Depth m
Clay, silty, stiff, moderate olive-brown (5Y4/4) mottled pale blue-green (5BG7/2)	0.20	1.50
Clay, silty, stiff, pale brown (5YR5/2) mottled medium yellow-brown (10YR4/6) with scattered carbonaceous material. Fissures infilled with coarse sand	1.60	3.10
Sand, fine-grained, clayey and silty clay, light brown (10YR5/6). Scattered selenite	1.40	4.50
Sand, fine-grained, more clayey than above. Light brown (10YR5/6). Scattered selenite crystals	0.80	5.30
Clay, silty, stiff, pale brown (5YR5/2) with clayey, fine sand lenses, moderate yellow brown (10YR5/4)	2.20	7.50
Clay, silty stiff, medium brown (5YR4/2) with sand lenses	0.10	7.60
Clay, silty, stiff, olive-black (5Y2/1). Laminated, scattered selenite crystals, fissured	seen to 1.90	9.50

APPENDIX 3

List of Geological Survey photographs

Copies of these photographs are deposited in the Libraries of the
Geological Museum, South Kensington, London, SW 7 2DE,
and of the British Geological Survey, Keyworth, Nottingham.
Prints and slides can be supplied at a standard tariff.
   All numbers belong to Series A.

2520   Valley of a small stream cut in London Clay, north of the
River Crouch

2521   Steep northern slope of the Crouch valley at Ramsden
Bellhouse with gentle slope of valley bottom. The flat is mainly
on London Clay

12246   Bagshot Beds: disused brickpit at Hambro Hill, Rayleigh
[TQ 8132 9191]

12247   Fossil ice-wedge polygons: Barling Hall Sand and Gravel
Pit, Barling Magna. The fill of ice-wedge polygons remains as
free standing vertical walls after removal of the surrounding
sand and gravel [TQ 9355 8168]

12248   Sand and gravel extraction from Crouch First Terrace:
Barling Sand and Gravel Pit, Barling Magna [TQ 9355 8168]

12249   Brickearth extraction: Cherry Orchard Lane Brickworks,
Rochford [TQ 8550 8950]

12250   Reclamation after Brickearth extraction: Cherry Orchard
Lane Brickworks, Rochford [TQ 8570 8940]

12252   Landslipping at Hadleigh. View of Hadleigh Castle built
above former sea cliffs of London Clay and Claygate Beds. The
landslips are considerably degraded and give rise to uneven
hummocky ground [TQ 809 858]

12253   View of Canvey Island and the Thames Estuary from
Sand Pit Hill, Hadleigh [TQ 801 862]

12254   BGS field laboratory for the South Essex Project; at
Great Stambridge Sewage Works [TQ 887 913]

12255   Crouch Second Terrace Deposits: Doggetts Pit, Rochford
[TQ 887 913]

12256   Crouch Second Terrace Deposits, showing involution
structures: Doggetts Pit, Rochford [TQ 887 913]

FOSSIL INDEX

Note: In compiling this index, signs which qualify identification (?, cf., aff., etc.) have been omitted.

GENERAL INDEX

Printed in the UK for HMSO 0738922.2M.0.00.48441

BRITISH GEOLOGICAL SURVEY

Keyworth, Nottinghamshire NG12 5GG

Murchison House, West Mains Road,
Edinburgh EH9 3LA

The full range of Survey publications is available
through the Sales Desks at Keyworth and
Murchison House. Selected items are stocked by
the Geological Museum Bookshop, Exhibition
Road, London SW7 2DE; all other items may be
obtained through the BGS London Information
Office in the Geological Museum. All the books
are listed in HMSO's Sectional List 45. Maps are
listed in the BGS Map Catalogue and Ordnance
Survey's Trade Catalogue. They can be bought
from Ordnance Survey Agents as well as from
BGS.

*On 1 January 1984 the Institute of Geological Sciences
was renamed the British Geological Survey. It continues to
carry out the geological survey of Great Britain and
Northern Ireland (the latter as an agency service for the
government of Northern Ireland), and of the surrounding
continental shelf, as well as its basic research projects. It
also undertakes programmes of British technical aid in
geology in developing countries as arranged by the Overseas
Development Administration.*

*The British Geological Survey is a component body of the
Natural Environment Research Council.*

HER MAJESTY'S STATIONERY OFFICE

HMSO publications are available from:

HMSO Publications Centre
(Mail and telephone orders)
PO Box 276, London SW8 5DT
Telephone orders (01) 622 3316
General enquiries (01) 211 5656

HMSO Bookshops
49 High Holborn, London WC1V 6HB
   (01) 211 5656 (Counter service only)
258 Broad Street, Birmingham B1 2HE
   (021) 643 3757
Southey House, 33 Wine Street, Bristol BS1 2BQ
   (0272) 24306/24307
9 Princess Street, Manchester M60 8AS
   (061) 834 7201
80 Chichester Street, Belfast BT1 4JY
   (0232) 238451
13a Castle Street, Edinburgh EH2 3AR
   (031) 225 6333

HMSO's Accredited Agents
(see Yellow Pages)

And through good booksellers